Lecture Notes in Mathematics

Edited by J.-M. Morel, F. Takens and B. Teissier

Editorial Policy
for the publication of monographs

1. Lecture Notes aim to report new developments in all areas of mathematics – quickly, informally and at a high level. Monograph manuscripts should be reasonably self-contained and rounded off. Thus they may, and often will, present not only results of the author but also related work by other people. They may be based on specialized lecture courses. Furthermore, the manuscripts should provide sufficient motivation, examples and applications. This clearly distinguishes Lecture Notes from journal articles or technical reports which normally are very concise. Articles intended for a journal but too long to be accepted by most journals, usually do not have this "lecture notes" character. For similar reasons it is unusual for doctoral theses to be accepted for the Lecture Notes series.

2. Manuscripts should be submitted (preferably in duplicate) either to one of the series editors or to Springer-Verlag, Heidelberg. In general, manuscripts will be sent out to 2 external referees for evaluation. If a decision cannot yet be reached on the basis of the first 2 reports, further referees may be contacted: the author will be informed of this. A final decision to publish can be made only on the basis of the complete manuscript, however a refereeing process leading to a preliminary decision can be based on a pre-final or incomplete manuscript. The strict minimum amount of material that will be considered should include a detailed outline describing the planned contents of each chapter, a bibliography and several sample chapters.
Authors should be aware that incomplete or insufficiently close to final manuscripts almost always result in longer refereeing times and nevertheless unclear referees' recommendations, making further refereeing of a final draft necessary.
Authors should also be aware that parallel submission of their manuscript to another publisher while under consideration for LNM will in general lead to immediate rejection.

3. Manuscripts should in general be submitted in English.
Final manuscripts should contain at least 100 pages of mathematical text and should include
 – a table of contents;
 – an informative introduction, with adequate motivation and perhaps some historical remarks: it should be accessible to a reader not intimately familiar with the topic treated;
 – a subject index: as a rule this is genuinely helpful for the reader.

Continued on inside back-cover

Lecture Notes in Mathematics 1784

Editors:
J.-M. Morel, Cachan
F. Takens, Groningen
B. Teissier, Paris

Subseries:
Fondazione C.I.M.E., Firenze
Adviser: Pietro Zecca

Springer
Berlin
Heidelberg
New York
Barcelona
Hong Kong
London
Milan
Paris
Tokyo

L. H. Eliasson   S. B. Kuksin
S. Marmi   J.-C. Yoccoz

Dynamical Systems and Small Divisors

Lectures given at the C.I.M.E. Summer School
held in Cetraro, Italy, June 13 - 20, 1998

Editors: S. Marmi
J.-C. Yoccoz

Fondazione
C.I.M.E.

Springer

Editors and Authors

Stefano Marmi
Scuola Normale Superiore
Piazza dei Cavalieri 7
56126 Pisa, Italy
e-mail: marmi@sns.it
and
Dipartimento di Matematica
e Informatica
Universitá di Udine
Via delle Scienze 206
33100 Udine, Italy
e-mail: marmi@dimi.uniud.it

Jean-Christophe Yoccoz
Collége de France
3 rue d' Ulm
75005 Paris, France
e-mail:
jean-c.yoccoz@college-de-france.fr

Hakan Eliasson
Géométrie et Dynamique
UFR. de Mathématiques
Université Paris 7
Case 7012, 2 place Jussieu
75251 Paris Cedex 05, France
e-mail: hakane@math.jussieu.fr

Sergei Kuksin
Department of Mathematics
Heriot-Watt University
Riccarton, Edinburgh EH14 4AS
United Kingdom
e-mail: s.b.kuksin@ma.hw.ac.uk

Cataloging-in-Publication Data applied for

Die Deutsche Bibliothek - CIP-Einheitsaufnahme

Dynamical systems and small divisors : held in Cetraro, Italy, Juny 13 - 20, 1998 / CIME. L. H. Eliasson ... Ed.: S. Marmi ; J.-C. Yoccoz. - Berlin ; Heidelberg ; New York ; Barcelona ; Hong Kong ; London ; Milan ; Paris ; Tokyo : Springer, 2002
    (Lectures given at the CIME summer school ; 1998, summer school)
    (Lecture notes in mathematics ; Vol. 1784 : Subseries: Fondazione CIME)
    ISBN 3-540-43726-6

Mathematics Subject Classification (2000):
37C55, 37F25, 37F50, 37J40, 37K55, 47B39, 34L40

ISSN 0075-8434
ISBN 3-540-43726-6 Springer-Verlag Berlin Heidelberg New York

Springer-Verlag Berlin Heidelberg New York a member of BertelsmannSpringer
Science + Business Media GmbH

http://www.springer.de

© Springer-Verlag Berlin Heidelberg 2002
Printed in Germany

Typesetting: Camera-ready TeX output by the authors

SPIN: 10878489   41/3142/du - 543210 - Printed on acid-free paper

Preface

The C.I.M.E. Session "Dynamical Systems and Small Divisors" was held in
Cetraro (Cosenza, Italy) from June 13 to June 20, 1998.

Lecture series were held by:

- L.H. Eliasson (8 hours): Linear quasiperiodic systems
- S. Kuksin (3 hours): KAM theory and PDEs
- M.R. Herman (8 hours): Abstract methods in small divisors: Implicit function theorems in Fréchet spaces
- J.N. Mather (3 hours): Variational construction of orbits
- J.-C. Yoccoz (8 hours): Geometrical methods in small divisor problems

In addition there was a 3 hours long open problem session.
Furthermore, the following participants gave a talk:

1. U. Bessi: A counterexample to KAM theorem
2. D. De Latte: Linearization of commuting holomorphic maps
3. G. Gentile: Scaling properties near resonances for the (semistandard and) standard map
4. H. Ito: Integrable symplectic maps and their Birkhoff normal forms
5. R. Krikorian: Reducibility of linear quasiperiodic systems: a global density result
6. P. Moussa: Regularity properties of Brjuno functions
7. D. Sauzin: Quasianalytic monogenic solutions of a cohomological equation
8. S. Smirnov: Weak expansion and geometry of Julia sets
9. L. Stolovitch: Singular complete integrability
10. M. Vittot: Perturbation method for Floquet hamiltonians with dense point spectrum

This volume contains expanded versions of the lecture series of Eliasson,
Kuksin and Yoccoz and of the open problem session.

Many problems of stability in the theory of dynamical systems face the
difficulty of small divisors. The most famous example is probably given by
Kolmogorov–Arnol'd–Moser theory on the persistence of quasi-periodic solutions of Hamilton's equations for quasi-integrable Hamiltonian systems (both
finite and infinite-dimensional, like nonlinear wave equations). This is a very
natural situation with many applications to physics and astronomy. What all
these different problems have in common is roughly speaking what follows:
one can associate some "frequencies" to the orbits under investigation and
some arithmetical condition is needed to prove their existence and stability. In

the smooth case one must impose a diophantine condition on the frequencies, in the analytic (finite–dimensional) case one can impose a weaker condition.

For one-dimensional analytic small divisor problems the optimal arithmetical condition to impose is now known: in the local case it is the same Brjuno condition introduced some 30 years ago, whereas in the global case one has to impose a more restrictive condition: the lectures by J.-C. Yoccoz published in these notes deal with these cases.

In his lectures Hakan Eliasson addresses the problem of reducibility of linear quasiperiodic skew-products. Originating from the pioneering works of Dinaburg–Sinai and Rüssmann, who constructed Floquet solutions for the continuous Schrödinger equation with a quasiperiodic potential in the socalled weak coupling regime (the potential is "small" w.r.t. the kinetic part), many authors have since then studied more general situations (strong coupling, discrete case, more general groups, etc.). The review of Eliasson at the ICM 1998 could serve as a general introduction to this topic.

The contribution of S. Kuksin is devoted to a proof of KAM theorem of persistence of finite–dimensional tori under small Hamiltonian perturbations of Lax-integrable Hamiltonian PDEs. Here the model problem is provided by Korteweg–de Vries equation $u_t = -u_{xxx} + 6uu_x$ where the spatial variable x varies on the one–dimensional torus. Also other problems can be treated by the same method: sine-Gordon equation, (φ^4)-equation, etc.. The study of nonlinear Hamiltonian PDEs is a rapidly growing research subject: excellent reviews of recent work are provided by the addresses of Bourgain at the ICM 1994 and Kuksin at the ICM 1998.

Finally, the open problems proposed by the speakers have been collected in the last contribution. They deal both with finite and infinite–dimensional small divisor problems. As far as possible we have tried to make this text self-contained.

We are grateful to all speakers and partecipants for their essential contribution to the success of this C.I.M.E. session. Special thanks go to Roberta Fabbri who helped us in the organization and to Raphael Krikorian who held several tutorials on the material of Herman's lectures.

Stefano Marmi and Jean-Christophe Yoccoz

Contents

Perturbations of linear quasi-periodic system 1
L. H. Eliasson
1 Introduction 1
2 Covariance and normal form matrices 13
3 Block splitting 17
4 Quadratic convergence 18
5 Block clustering 25
6 Transversality of resultants 29
7 A Perturbation theorem 34
8 Applications 42
A Appendix A 46
B Appendix B 56

KAM-persistence of finite-gap solutions 61
Sergei B. Kuksin
Introduction 61
1 Some analysis in Hilbert spaces and scales 61
2 Symplectic structures and Hamiltonian equations 67
3 Lax-integrable Hamiltonian equations
   and their integrable subsystems 74
4 Finite-gap manifolds and theta-formulas 80
5 Linearised equations and their Floquet solutions 86
6 Linearised Lax-integrable equations 96
7 Normal form 104
8 The KAM theorem 111
9 Examples 117

Analytic linearization of circle diffeomorphisms 125
Jean-Christophe Yoccoz
1 Introduction 125
2 Arithmetics 127
3 The C^r theory for $r \geq 0$ 137
4 Analytic case 146
5 Appendix: Estimates of moduli of annular domains 168

Some open problems related to small divisors 175
S. Marmi, J.-C. Yoccoz

0   Introduction 175
1   One-Dimensional Small Divisor Problems (On Holomorphic Germs
    and Circle Diffeomorphisms) 175
2   Finite-Dimensional Small Divisor Problems 178
3   KAM Theory and Hamiltonian Systems 181
4   Linear Quasiperiodic Skew-Products, Spectral Theory and Hamil-
    tonian Partial Differential Equations 184

Perturbations of linear quasi-periodic system

L. H. Eliasson

1   Introduction

Existence of both Floquet and l^2 solutions of linear quasi-periodic skew-products can be formulated in terms of linear operators on $l^2(\mathbb{Z})$, i.e. ∞-dimensional matrices. In the perturbative regime these matrices are perturbations of diagonal matrices and the problem is to diagonalize them completely or partially, i.e. to show that they have some *point spectrum*.

The unperturbed matrices have a *dense point spectrum* so that their eigenvalues are, up to any order of approximation, of *infinite multiplicity*, which is a very delicate situation to perturb. For matrices with *strong decay* of the matrix elements off the diagonal this difficulty can be overcome if the eigenvectors are sufficiently well *clustering*. One way to handle this is to control the *almost multiplicites* of the eigenvalues.

The eigenvalues are given by functions of one or several parameters and in order to control the almost multiplicities it is necessary that these functions are not too flat. Such a condition is delicate to verify since derivatives of eigenvalues of a matrix behave very bad under perturbations of the matrix. Derivatives of eigenvalues of matrices are therefore replaced by derivatives of *resultants of matrices* – an object which behaves better under perturbations.

If the parameter space is *one-dimensional* and if the quasi-periodic frequencies satisfy some *Diophantine condition*, then it turns out that this control of the derivatives of eigenvalues, in terms of the resultants, is not only necessary but also sufficient for the control of the almost multiplicities. If the parameter space is higher-dimensional this control is more difficult to achieve and not yet well understood.

1.1   Examples

Discrete Schrödinger equations in one dimension. In *strong coupling* regime this equation takes the form

$$-\varepsilon(u_{n+1} + u_{n-1}) + V(\theta + n\omega)u_n = Eu_n, \qquad n \in \mathbb{Z}, \qquad (1.1)$$

where V is a real-valued piecewise smooth function on the d-dimensional torus $\mathbb{T}^d = (\mathbb{R}/2\pi\mathbb{Z})^d$ and ω is a vector in $\mathbb{R}^d$. The constant ε is a way to measure the size of the potential V and it is assumed to be small, i.e. a large potential. E is a real parameter.

When $\varepsilon = 0$ (1.1) have for each $m \in \mathbb{Z}$ the solution

$$u^m \in l^2(\mathbb{Z}), \qquad u_n^m = e_n^m$$

if we choose $E = V(\theta + m\omega) - e_n^m$ is 1 when $n = m$ and 0 otherwise.

A solution $u \in l^2(\mathbb{Z})$ is an eigenvector of the left-hand side of (1.1) interpreted as an operator on $l^2(\mathbb{Z})$. Represented in the standard basis for $l^2(\mathbb{Z})$ it becomes an ∞-dimensional matrix

$$\begin{pmatrix} \ddots & & \\ & V(\theta + n\omega) & \\ & & \ddots \end{pmatrix} + \varepsilon \begin{pmatrix} \ddots & & & & \\ & 0 & -1 & & \\ & -1 & 0 & -1 & \\ & & -1 & 0 & \\ & & & & \ddots \end{pmatrix}$$

over $\mathbb{Z}$. This matrix has the following properties:

* *dependence of parameters* $\theta \in \mathbb{T}^d$;
* *covariance* with respect to the $\mathbb{Z}$-action on $\mathbb{T}^d$

$$\theta \mapsto \theta + n\omega$$

– covariance means that we obtain any row/column, and hence the whole matrix, from one row/column simply by shifting the parameter θ by the group action;

* *dense spectrum* of the diagonal part;
* *domination of diagonal part* when ε is small;
* *strong decay off the diagonal*.

We can also notice that the matrix is *symmetric* and that the diagonal part is determined by the potential V while the perturbation is independent of the potential.

A natural question is if this matrix has a basis of eigenvectors in $l^2(\mathbb{Z})$, in which case it can be diagonalized, or if it has any eigenvectors at all. Such eigenvectors will give solutions to (1.1) which are decaying both in forward and backward time.

The existence of eigenvectors for small but positive ε is a delicate question because in the unperturbed case all eigenvalues are essentially of infinite multiplicity. Moreover there are numerous examples when there are no eigenvectors: if V is constant there are no eigenvectors since the operator $u_n \to -\varepsilon(u_{n+1} + u_{n-1})$ is absolutely continuous on $l^2(\mathbb{Z})$; if the potential is periodic, i.e. $\omega \in 2\pi\mathbb{Q}^d$, there are no eigenvectors; if the potential is nonperiodic there are examples without eigenvectors when ω is Liouville.

In *weak coupling* we consider the equation in the form

$$-(u_{n+1} + u_{n-1}) + \varepsilon V(\theta + n\omega)u_n = Eu_n, \qquad n \in \mathbb{Z}. \qquad (1.2)$$

When $\varepsilon = 0$ (1.2) has, for each $\xi \in \mathbb{T}$, extended solutions

$$u^\xi \in l^\infty(\mathbb{Z}), \qquad u_n^\xi = e^{in\xi},$$

for $E = -2\cos(\xi)$. When ε is small it is natural to look for extended solutions of the form

$$u_n^\xi = e^{in\xi}U(\theta + n\omega),$$

where $U : \mathbb{T}^d \to \mathbb{C}$. Such solutions are known as *Floquet solutions* or *Bloch waves*. The equation for U then becomes

$$-(e^{i\xi}U(\theta + \omega) + e^{-i\xi}U(\theta - \omega)) + \varepsilon V(\theta)U(\theta) = EU(\theta)$$

which, when written in Fourier coefficients, gives the matrix

$$
\begin{pmatrix}
\ddots & & \\
& 2\cos(\xi + <k,\omega>) & \\
& & \ddots
\end{pmatrix}
- \varepsilon
\begin{pmatrix}
\ddots & & \\
& 0 & \hat{V}(k-l) \\
& \hat{V}(l-k) & 0 \\
& & & \ddots
\end{pmatrix}
$$

over $\mathbb{Z}^d$. This matrix has the properties:

* *dependence of parameters $\xi \in \mathbb{T}$;*
* *covariance with respect to the $\mathbb{Z}^d$-action on $\mathbb{T}$*

$$\xi \mapsto \xi + <k,\omega>.$$

It has like (1.1) a *dense spectrum* of the diagonal part, *domination of diagonal part* when ε is small and *decay off the diagonal* (depending on the smoothness of V). In this case however, the matrix is complex and *Hermitian*, the diagonal part is fixed $= 2\cos(\alpha)$ and the perturbation depends on the potential function V.

Finding Floquet solutions for (1.2) now amounts to finding eigenvectors of this matrix. This is a delicate matter which is known to depends on arithmetical properties of ω and requires strong smoothness of V which is reflected in strong decay off the diagonal of the matrix. Floquet solutions are related to the absolutely continuous spectrum and there are examples of discontinuous potentials with only singular spectrum.

In strong coupling we can consider generalizations of the form

$$-\varepsilon(\Delta_\theta u)_n + V(\theta + n\omega)u_n = Eu_n, \qquad n \in \mathbb{Z}, \tag{1.1'}$$

where the difference operator is allowed to take a more general form

$$(\Delta_\theta u)_n = \sum_{j=-N}^{N} b_j(\theta + (n+j)\omega)u_{n+j}.$$

It could even be infinite with sufficiently strong decay.

This equation will now give rise to a perturbation problem with the same diagonal part as (1.1) but with a perturbation that depends on the b_j's. It is typically not symmetric and it is therefore not reasonable to try to diagonalize it for small ε. But block diagonalization and construction of eigenvectors still make sense.

In weak coupling we can consider equation

$$-(\Delta u)_n + \varepsilon V(\theta + n\omega)u_n = Eu_n, \qquad n \in \mathbb{Z}, \qquad (1.2')$$

with a constant difference operator of the more general form

$$(\Delta u)_n = \sum_{j=-N}^{N} b_j u_{n+j}.$$

It can also be infinite with sufficiently strong decay. It will now give rise to a perturbation problem of the same type as (1.2) but with diagonal part determined by the function

$$\sum_{j=-N}^{N} b_j e^{ij\xi}.$$

Discrete Linear Skew-Products. We consider a *weakly perturbed* skew-product of the form

$$X_{n+1} - (A + \varepsilon B(\theta + n\omega))X_n = EX_n, \qquad n \in \mathbb{Z}, \qquad (1.3)$$

where $A \in Gl(N, \mathbb{R})$, $B : \mathbb{T}^d \to gl(N, \mathbb{R})$.

When $\varepsilon = 0$ we have the Floquet solutions

$$X_n = e^{in\xi}X_0, \qquad AX_0 = aX_0,$$

for $E = e^{i\xi} - a$, $\xi \in \mathbb{T}$. For $\varepsilon \neq 0$ we look for Floquet solutions

$$X_n = e^{in\xi}Y(\theta + n\omega)$$

with $Y : \mathbb{T}^d \to \mathbb{C}^N$. The equation for Y then becomes

$$e^{i\xi}Y(\theta + \omega) - (A + \varepsilon B(\theta))Y(\theta) = EY(\theta)$$

which, when expressed in Fourier coefficients, becomes the matrix

$$\begin{pmatrix} \ddots & & \\ & e^{i(\xi + <k,\omega>)}I - A & \\ & & \ddots \end{pmatrix} - \varepsilon \begin{pmatrix} \ddots & & \\ & 0 & \hat{B}(k-l) \\ & \hat{B}(l-k) & 0 \\ & & \ddots \end{pmatrix}$$

over $\mathbb{Z}^d$. The properties of the matrix are:

* *dependence of parameters* $\xi \in \mathbb{T}$;
* *covariance* with respect to the $\mathbb{Z}^d$-action on $\mathbb{T}$

$$\xi \mapsto \xi + <k, \omega>.$$

The matrix has *dense point spectrum* of diagonal part, *domination of diagonal part* when ε is small and *decay off the diagonal* (depending on the smoothness of B). Notice that here the unperturbed part is only block diagonal and its spectrum is determined by the functions $(e^\xi - a_j)$ where a_j runs over the eigenvalues of A – a *multi-level operator*.

Finding Floquet solutions for (1.3) now amounts to finding eigenvectors of this matrix.

We shall give two example of skew-products which can be considered as *strongly perturbed*. The first example is an obvious generalization of the Schrödinger equation:

$$-\varepsilon(X_{n+1} + X_{n-1}) + B(\theta + n\omega)X_n = EX_n, \qquad n \in \mathbb{Z}, \qquad (1.4)$$

where $B : \mathbb{T}^d \to gl(N, \mathbb{R})$ is a symmetric matrix. This now gives rise to a multi-level version of (1.1).

The second example arises from the continuous Schrödinger equation in strong coupling

$$-\varepsilon u''(t) + V(\theta + t\omega)u(t) = Eu(t), \qquad t \in \mathbb{R}. \qquad (1.5)$$

For $\varepsilon = 0$ the operator to the left becomes multiplication by $V(\theta + t\omega)$ which has purely continuous spectrum and no eigenvalues. Hence this is an even more complicated object to perturb from. Instead we consider the corresponding dynamical system

$$\begin{pmatrix} u \\ v \end{pmatrix}' = \begin{pmatrix} 0 & 1 \\ \frac{1}{\varepsilon}(V(\theta + t\omega) - E) & 0 \end{pmatrix} \begin{pmatrix} u \\ v \end{pmatrix}, \qquad \theta' = \theta + \omega,$$

and its time-t-map $(\Phi_t(\theta; E, \varepsilon), \theta + \omega)$.

Let $\omega = (\omega', 1)$ and $\theta = (\theta', \theta_d)$. Then the evolution of the system is described by

$$\begin{pmatrix} u_{n+1} \\ v_{n+1} \end{pmatrix} = \begin{pmatrix} a_n & b_n \\ c_n & d_n \end{pmatrix} \begin{pmatrix} u_n \\ v_n \end{pmatrix} =: \Phi_1((\theta' + n\omega', 0); E, \varepsilon) \begin{pmatrix} u_n \\ v_n \end{pmatrix},$$

where $a_n d_n - b_n c_n = 1$. From this we deduce the second order equation

$$b_{n-1}u_{n+1} + b_n u_{n-1} - W(\theta' + n\omega')u_n = 0, \qquad (1.6)$$

where $W(\theta') = a_0(\theta')b_0(\theta' - \omega') + d(\theta' - \omega')b(\theta')$.

In matrix formulation the right-hand side of (1.6) becomes

$$\begin{pmatrix} \ddots & & \\ & W(\theta' + n\omega') & \\ & & \ddots \end{pmatrix} + \begin{pmatrix} \ddots & & & \\ & 0 & b_{n-2} & \\ & b_n & 0 & b_{n-1} \\ & & b_{n+1} & 0 \\ & & & & \ddots \end{pmatrix}.$$

which is a matrix over $\mathbb{Z}$ with properties:

* *dependence of parameters* $\theta' \in \mathbb{T}^{d-1}$;
* *covariance* with respect to the $\mathbb{Z}$-action on $\mathbb{T}^{d-1}$

$$\theta' \mapsto \theta' + n\omega'.$$

The diagonal part has *dense spectrum* and the matrix has *strong decay off the diagonal*. There is *some domination of the diagonal part* if ε is small and if E is close to $\inf V$ but it is less explicit than in (1.1). Both the diagonal and the non-diagonal part depend on the potential V.

Since the matrix is neither symmetric nor Hermitian we cannot hope to diagonalize it, but it may still be block diagonalizable and have a pure point spectrum with finite multiplicity. If 0 is an eigenvalue for a given parameter value E, then the corresponding eigenvector will give a solution of (1.5) which is decaying both in forward and backward time giving some point spectrum for that operator.

Higher-dimensional Schrödinger equations. We can consider the Schrödinger equation in, say, two dimensions in *strong coupling*

$$-\varepsilon(\Delta u)_n + V(\theta_1 + n_1\omega_1, \theta_2 + n_2\omega_2)u_n = Eu_n, \qquad n \in \mathbb{Z}^2, \qquad (1.7)$$

where Δ is the nearest neighbor difference operator

$$(\Delta u)_n = \sum_{|m-n|=1} u_m,$$

V is a real valued piecewise smooth function on the product of two tori $\mathbb{T}^d \times \mathbb{T}^d$ and $\omega = (\omega_1, \omega_2)$ is a vector in $\mathbb{R}^d \times \mathbb{R}^d$.

A solution $u \in l^2(\mathbb{Z}^2)$ is an eigenvector of a perturbation of the matrix

$$\begin{pmatrix} \ddots & & \\ & V(\theta_1 + n_1\omega_1, \theta_2 + n_2\omega_2) & \\ & & \ddots \end{pmatrix}$$

over $\mathbb{Z}^2$. This matrix has the following properties:

* *dependence of parameters* $\theta \in \mathbb{T}^{2d}$;
* *covariance* with respect to the $\mathbb{Z}^2$-action on $\mathbb{T}^{2d}$

$$\theta = (\theta_1, \theta_2) \mapsto (\theta_1 + n_1\omega_1, \theta_2 + n_2\omega_2).$$

The diagonal part has *dense spectrum* and *dominates* the matrix when ε is small, and *the decay off the diagonal is strong*.

In *weak coupling*

$$-(\Delta u)_n + \varepsilon V(\theta_1 + n_1\omega_1, \theta_2 + n_2\omega_2)u_n = Eu_n, \qquad n \in \mathbb{Z}^2, \qquad (1.8)$$

the existence of Floquet solutions

$$u_n = e^{i(n_1\xi_1 + n_2\xi_2)}U(\theta_1 + n_1\omega_1, \theta_2 + n_2\omega_2)$$

gives rise to an eigenvalue problem in the same way as (1.2).

1.2 Two basic difficulties

The fundamental problem in the perturbation theory of these matrices is that all eigenvalues have infinite multiplicity up to any order of approximation. More precisely, we say that two eigenvalues $E^m(\theta)$ and $E^n(\theta)$ are ρ-*almost multiple* if

$$|E^m(\theta) - E^n(\theta)| \leq \rho.$$

Then the ρ-almost multiplicity is infinite for any $\rho \neq 0$. This is not only a technical obstacle since for example the matrix derived from equation (1.1) does not have any eigenvectors for any $\varepsilon \neq 0$ if V is constant.

Infinite almost-multiplicity can be handled if the matrix has strong decay off the diagonal and if the corresponding almost-multiple eigenvectors $\{u^{m_i}\}$ cluster. By this we mean that the range of all the (u^{m_i})'s, i.e. the infinite subset

$$\cup_i\{n \in \mathbb{Z} : u_n^{m_i} \neq 0\}$$

collects into finite clusters in $\mathbb{Z}$. This clustering of almost-multiple eigenvectors is crucial for the possible diagonalization of the perturbed matrix. It should be measured by the *separation* and *extension* of the clusters – which are related to the decay of the matrix off the diagonal – and by the the *cardinality* of the clusters.

The first difficulty is therefore to get good clustering of eigenvectors. An essential role in this is played by the *strong decay* of the perturbation off the diagonal and by the *Diophantine properties* of the frequency vector ω. A vector $\omega \in \mathbb{R}^d$ is said to be *Diophantine* with parameters $\kappa, \tau > 0$ if

$$\inf_{m \in \mathbb{Z}} |<n, \omega> -2\pi m| \geq \frac{\kappa}{|n|^{\tau}}, \qquad \forall n \in \mathbb{Z}^d \setminus 0. \qquad (1.9)$$

Control of the almost multipicities and clustering. In the problem (1.1) the eigenvalues $E^m(\theta)$ and the eigenvectors $u^m(\theta)$ are

$$E^m(\theta) = V(\theta + m\omega) \quad \text{and} \quad u_n^m = e_n^m.$$

Which are the eigenvalues that are ρ-almost multiple to $E^m(\theta)$ and how many are there in any finite subset of $\mathbb{Z}$?

In order to investigate this question, consider the set

$$\Delta = \{x \in \mathbb{T} :\mid E^m(\theta + x) - E^m(\theta) \mid \leq \rho\}.$$

If V is *smooth*, measured by β and γ,

$$\mid V \mid_{C^k} \leq \beta (k!)^2 \gamma^k \qquad \forall k \geq 0,$$

and if V satisfies the *transversality condition*, measured by σ and s,

$$\max_{0 \leq k \leq s} \mid \partial_x^k (V(\theta + x) - V(\theta)) \mid \geq \sigma \qquad \forall \theta, x,$$

then Δ (see Lemma B1) is a union of intervals $\cup \Delta_j$ with a finite number $\leq \mu$ (independent of ε) of components, and each component is contained in an interval with length

$$\leq \text{const } \rho^{\frac{1}{s}}$$

– when $V(\theta) = \cos(\theta)$ the numbers μ and s are both 2. If there are more than $\mu + 1$ eigenvalues $E^{n_j}(\theta)$, $j = 1, 2, \ldots, \mu + 1$, that are ρ-almost multiple to $E^m(\theta)$, then it follows that $\theta + n_i\omega$ and $\theta + n_j\omega$ belong to one component, Δ_1 say, for at least two different $i, j \leq \mu + 1$. What controls the return of a translation $\theta \to \theta + \omega$ to an interval is the Diophantine condition (1.9). Under this condition it follows that

$$\mid n_i - n_j \mid \geq \text{const} \kappa^{\frac{1}{\tau}} (\frac{1}{\rho})^{\frac{1}{\tau s}},$$

which means that when ρ is small then the distance between the sites n_i and n_j is very large.

Consider now a sequence of eigenvalues $E^m(\theta) = E^{n_1}(\theta), E^{n_2}(\theta), \ldots,$ $E^{n_{\mu+1}}(\theta)$ such that any two consecutive pairs are ρ-close. Then any two pairs are $(\mu + 1)\rho$-close and hence it must hold that

$$\max_{1 \leq i, j \leq \mu+1} \mid n_i - n_j \mid \geq \nu = \text{const} \kappa^{\frac{1}{\tau}} (\frac{1}{(\mu + 1)\rho})^{\frac{1}{\tau s}}.$$

This implies that the ranges of the corresponding eigenvectors u^{n_i}, i.e. the set of indices or sites $\{n_i\}$, collects into clusters in $\mathbb{Z}$ containing at most μ many sites. Any two clusters are separated by at least ν many sites, and each cluster extends over at most $\lambda = \mu\nu$ sites.

This simple argument which controls the clustering of the eigenvectors does not depend on the dimension of the lattice $\mathbb{Z}$, but it does depend crucially on the dimension of the parameter space $\{\theta\}$. If we consider $V(\theta) = \cos(\theta_1) + \cos(\theta_2)$, defined on the 2-dimensional torus, with $\omega = (\omega_1, \omega_2)$, then the corresponding set Δ is now a subset of $\mathbb{T}^2$ and its components Δ_i are simply connected subsets of $\mathbb{T}^2$ whose shape depend on the potential V. A Diophantine condition on ω does not alone control the return of a translation to an open connected set – the return also depend on the geometry of the sets Δ_i. Of course the situation is the same if we consider a truly higher-dimensional problem like (1.7) or (1.8).

This is why the higher-dimensional case is more difficult and why our results in this paper are restricted to matrices which are covariant with respect to a group action on the one-dimensional space $\mathbb{T}^1$.

Propagation of smoothness and transversality. The second difficulty is that the diagonalization is a *small divisor problem* which requires a quadratic iteration of KAM-type. It is therefore not enough to control the almost-multiplicities of the unperturbed matrix but they must be controlled also for nearby matrices – if not for all so at least for a sufficiently large class of nearby matrices. This class of matrices is the *normal form matrices*: their elements have exponential decay off the diagonal measured by a parameter α; their smoothness is measured by β, γ; they are not smooth on the whole parameter space but only piecewise smooth over a partition $\mathcal{P}$. They will have a pure point spectrum and will be characterized by a certain *block structure* – measured by λ, μ, ν, ρ – which describes their eigenvectors. However, they will not be diagonal nor block diagonal in general.

Since derivatives of eigenvalues behave very bad under perturbations we cannot hope that the transversality condition of the eigenvalues of the unperturbed matrix propagates to the eigenvalues of the nearby normal form matrices. But instead of comparing two eigenvalues we can compare the whole spectrum of two submatrices truncated over blocks. This is done by the resultant and what we need is a transversality condition of the resultants of submatrices truncated over blocks – *transversality of blocks*. In distinction to eigenvalues, derivatives of resultants behave nicely under perturbations and we are able to show that this transversality condition of a normal form matrix propagates to nearby normal form matrices *if* their block structures are compatible in a certain sense.

1.3 Results and References

Some attempts to prove pure point spectrum for the discrete quasi-periodic Schrödinger equation (1.1) *on a one-dimensional torus* in strong coupling were made already in the early 80's [1,2] but the breakthrough came in the late 80's with the work of Sinai [3] and of Fröhlich, Spencer and Wittver [4] on a "cosine"-like potential. The most general result today is [5]. The present article provides a different presentation of the result in [5], together with generalizations and variants. Similar results have also been obtained in [6] by different methods. Discrete equations (1.1′) were treated in [5,7]. In this paper we generalize the these results still a bit more.

It should be noted the we only discuss the perturbation theory for smooth systems with Diophantine frequencies. There are numerous results for systems that are non-smooth or have non-Diophantine frequencies. For example, the results referred to above about Liouville frequencies and discontinuous potentials can be found in [8,9].

The results on weak coupling (1.2) are older. Dinaburg and Sinai [10] constructed Floquet solutions for the continuous Schrödinger in order to get some absolutely continuous spectrum (see also [11]). These results were then extended in [12] and the proof that there is such a solution for almost all quasi-momentum was given in [13]. The work [13] even proves that the one-parameter family (1.2) have Floquet solutions for a.e. value of the spectral parameter E. All these results were proved for the continuous equation but the adaptation to the discrete case is likely to be quite straight forward. The results of [10] was for example carried over to the discrete case in [14]. The results [10–14] were proven by ODE-methods. An operator approach like the one in this paper were used in [15] and later also in [16].

For more general weakly perturbed skew-products (1.3) results have been obtained in [17–19], also with the use ODE-methods. (See also the short survey [20].) Some further results are obtained in this paper.

The continuous Schrödinger in strong coupling (1.5) has only been treated in one particular case [4]:

$$-\varepsilon u''(t) \; + \; (cos(\theta_1 + t) + cos(\theta_2 + t\omega))u(t) = Eu(t).$$

This equation has, for ε small enough, a pure point spectrum close to the bottom of the spectrum. (From [13] it follows that it has purely absolutely continuous spectrum with Floquet solutions for sufficiently large E.)

Other strongly perturbed skew-products have been studied in [21] from the point of view of Lyapunov exponents.

For equation (1.1) *on a higher-dimensional torus* there are two papers [22,23] which, in particular, stresses the importance of "external" parameters in these problems. In truly higher dimension essentially nothing is proven. For the continuous analogue of (1.8) some results, also with the use of "external" parameters, have been announced [24,25]. The difficulties met in higher dimensions are related to those in other problems on infinite-dimensional matrices [26].

In this paper we only discuss discrete quasi-periodic skew-products. The continuous case is both similar and different. The construction of Floquet solution does give rise to a similar perturbation problem. The difference is that the parameter space is the Euclidean space $\mathbb{R}$, instead of the compact circle $\mathbb{T}$, and that the spectrum of the diagonal part is unbounded. This is likely to be a minor point and we believe that all the results in weak coupling carry over from discrete to continuous skew-products. The construction of l^2 solutions of continuous skew-products is, however, much more difficult since it does not give rise to such a nice perturbation problem as in the discrete case.

1.4   Description of the paper

In Section 1 we introduce some notations. In Section 2 we define the normal form matrices which are a sort of generalized block matrices. They depend smoothly on some parameters and are covariant with respect to a group action. The smoothness is only piecewise in a way we explain. In Section 3 we show that we can conjugate the normal form matrices to block diagonal form, but there is a prize to pay – the smoothness properties of the conjugated matrix is less good. In Section 4 we show that we can conjugate a perturbation of a normal form matrix to a new normal form matrix with a much smaller perturbation. This should be the starting point for an iteration of KAM-type. In Section 5 we discuss under what conditions this procedure can be iterated and the role of clustering of the blocks. Up to this point neither the smoothness nor the group action play any role whatsoever.

The clustering property is related to almost multiplicities of the eigenvalues. These multiplicities are related to estimates of resultants which can be obtained from a transversality property. In Section 6 we discuss when and how this transversality property of the resultants can be transferred to nearby normal form matrices.

In Section 7 we specialize to a quasi-periodic group action on the one-dimensional torus, i.e. to a linear ergodic action on $\mathbb{T}$, submitted to a Diophantine condition. In this case the transversality property of the resultants will be sufficient to control the almost multiplicities of the eigenvalues and hence the clustering of the blocks. This will give us a perturbation theorem, Theorem 12, of a quite general type. In Section 8 we discuss applications of Theorem 12 to some of the one-dimensional problems we have discussed above.

In an appendix we include basic finite-dimensional results from analysis and linear algebra.

There is some difference, besides presentation, from the approach in [5]. We work with Gevrey functions – the particular Gevrey class has no importance – but instead of using the possibility to smoothly truncate Gevrey function as in [5] we use here discontinuous cut-offs. The disadvantage is that we have to work with piecewise smooth functions, which is a little awkward in relation to the covariance property, and that we must control the size of the pieces. The advantage is that the particular group action does not intervene before the final step of the proof.

1.5   Notations

Let $\mathcal{L}$ be a standard lattice $\mathbb{Z}, \mathbb{Z}^2, \mathbb{Z}^3, \dots$ and denote its elements, called *indices* or *sites*, by $a, b, c, \dots$. If $\Omega, \Omega' \subset \mathcal{L}$, then we let

$$\underline{\mathrm{dist}}(\Omega, \Omega') = \min_{a \in \Omega, b \in \Omega'} |a - b|,$$

where $|\ \ |$ denotes the Euclidean distance. We let $\{e^b : b \in \mathcal{L}\}$ denote the standard basis of $l^2(\mathcal{L})$, defined by $e_a^b = 1$ or 0 depending on if $a = b$ or not. If Ω is a subset of $\mathcal{L}$, we denote by $\mathbb{C}^\Omega$ the subspace of $l^2(\mathcal{L})$ spanned by $\{e^a : a \in \Omega\}$.

A *subspace* Λ of dimension k of $l^2(\mathcal{L})$ comes equipped with an ON-frame, i.e. Λ is a linear mapping

$$\Lambda : \mathbb{C}^k \to l^2(\mathcal{L})$$

such that

$$\begin{cases} \Lambda^*\Lambda = Id_{\mathbb{C}^k} \\ \Lambda\Lambda^* = \perp - \text{projection of } l^2(\mathcal{L}) \text{ onto } \Lambda. \end{cases}$$

The *range* of Λ is the set of sites that Λ occupies, i.e. the set of all $a \in \mathcal{L}$ such that $(e^a)^*\Lambda \neq 0$. A *block* for Λ is a subset $\Omega \subset \mathcal{L}$ that contains the range of Λ.

If $D : l^2(\mathcal{L}) \to l^2(\mathcal{L})$ is a bounded operator we identify it with an ∞-dimensional matrix via the standard basis:

$$D_a^b = (e^a)^* D e^b,$$

the matrix element at row a and column b. By a *matrix* on $\mathcal{L}$ we shall always understand such a representation of a bounded operator. The *truncation at distance N off the diagonal* of D is the matrix $\tilde{D}$ defined by $\tilde{D}_a^b = D_a^b$ or $= 0$ depending on if $|b - a| \leq N$ or not. If Λ is a subspace we let $D_\Lambda = \Lambda^* D \Lambda$ and if $\Omega \subset \mathcal{L}$ we denote by D_Ω the matrix $D_{\mathbb{C}^\Omega}$.

A *partition* of a manifold X is a (locally finite) collection of open subsets – *pieces* – $\mathcal{P}$ such that $\cup_{Y \in \mathcal{P}} Y$ is of full measure in X. If $\mathcal{P}$ and $\mathcal{Q}$ are two such decomposition and if T is a homeomorphism on X, then

$$\begin{cases} \mathcal{P} \vee \mathcal{Q} = \{Y \cap Z : Y \in \mathcal{P}, Z \in \mathcal{Q}\} \\ T(\mathcal{P}) = \{T(Y) : Y \in \mathcal{P}\}. \end{cases}$$

$\mathbb{T}^d$ denotes the d-dimensional torus $(\mathbb{R}/2\pi\mathbb{Z})^d$ and its distance is denoted by $\|\ \ \|$.

We denote by $|u|$ the *Euclidean norm* if u is a vector and the *operator norm* if u is a linear operator.

For a smooth function u (possibly vector or matrix valued) defined in an open box V in $\mathbb{R}^d$ we let $|u|_{C^k}$ be

$$\sup_{0 \leq l \leq k} \sup_{x \in V} |\frac{1}{(l!)^2} \partial^l u_j(x)|, \qquad k \in \mathbb{N}^d.$$

u is said to be of *Gevrey class* $\mathcal{G}_2$ if

$$|u|_{C^k} \leq \beta\gamma^k \qquad \forall k \geq 0.$$

Smooth function will always be assumed to be $\mathcal{G}_2$ unless otherwise specified – that we have chosen this particular Gevrey class will be of no importance.

A function is said to be *smooth on* $\mathcal{P}$ if it is smooth on (a neighborhood of the closure of) each piece of $\mathcal{P}$.

2   Covariance and normal form matrices

2.1   Covariance

Let $\mathcal{L}$ be a lattice $\mathbb{Z}, \mathbb{Z}^2, \ldots$ and let X be a a torus $\mathbb{T}, \mathbb{T}^2, \ldots$. On the torus X we have an $\mathcal{L}$-action

$$T : \mathcal{L} \times X \to X,$$

and we denote by T_a the mapping $x \to T(a, x)$, $a \in \mathcal{L}$, $x \in X$. For an element b of $\mathcal{L}$, define $\tau_b : l^2(\mathcal{L}) \to l^2(\mathcal{L})$ by

$$(\tau_b f)_a = f_{a-b}.$$

Clearly τ_b is unitary and $\tau_b^* = \tau_b^{-1} = \tau_{-b}$.

Definition. *A matrix*

$$D : l^2(\mathcal{L}) \times X \to l^2(\mathcal{L})$$

is covariant with respect to the group action T if

$$\tau_c^* D(x) \tau_c = D(T_c(x)) \quad \forall c \in \mathcal{L}, \ x \in X.$$

A covariant matrix $D(x)$ is pure point if, for almost all x, there exists an eigenvector $q(x)$ such that $\{q^a(x) =: \tau_a q(T_a(x)) : a \in \mathcal{L}\}$ is a basis for $l^2(\mathcal{L})$. $\Omega(x) \subset \mathcal{L}$ is a block for $q(x)$ if

$$\Omega(x) \supset \{a \in \mathcal{L} : (e^a)^* q(x) \neq 0\}.$$

If $q(x)$ is an eigenvector with eigenvalue $E(x)$ and block $\Omega(x)$, $q^b(x) = \tau_b q(T_b(x))$ is an eigenvector with eigenvalue $E^b(x) = E(T_b(x))$ and block $\Omega^b(x) = \Omega(T_b(x)) - b$.

In terms of the matrix elements, covariance means that

$$D_{a+c}^{b+c}(x) = D_a^b(T_c x) \quad \forall a, b, c \in \mathcal{L}, \ x \in X.$$

More generally, we say that $D(x)$ is pure point if $q(x)$ is a generalized eigenvector with finite multiplicity. A block for $q(x)$ is then a block for the smallest invariant space containing $q(x)$.

Even more generally we can consider *multi-level operators*

$$D : l^2(\mathcal{L}) \otimes \mathbb{C}^N \times X \to l^2(\mathcal{L}) \otimes \mathbb{C}^N.$$

On $l^2(\mathcal{L}) \otimes \mathbb{C}^N$ we have the standard basis $\{e^{b,j} : b \in \mathcal{L}, \ j = 1, \ldots, N\}$ defined by $e^{b,j}_{a,i} = 1$ or 0 depending on if $(a, i) = (b, j)$ or not. For an element $b \in \mathcal{L}$, we define $\tau_b : l^2(\mathcal{L}) \otimes \mathbb{C}^N \to l^2(\mathcal{L}) \otimes \mathbb{C}^N$ by

$$(\tau_b f)_{a,i} = f_{a-b,i}, \qquad i = 1, \ldots, N.$$

The covariance of D with respect to the $\mathcal{L}$-action T is now defined in the same way as above. $D(x)$ is pure point if $l^2(\mathcal{L}) \otimes \mathbb{C}^N$ has a basis $\{q^{b,j}(x) : a \in \mathcal{L}, \ j = 1, \ldots N\}$ of eigenvectors of $D(x)$. A block for q^j is a subset $\Omega(x) \subset \mathcal{L} \times \{1, \ldots, N\}$ such that

$$\Omega(x) \supset \{(a, i) \in \mathcal{L} \times \{1, \ldots, N\} : (e^{a,i})^* q^j(x) \neq 0\}.$$

We can also consider in the multi-level case generalized eigenvectors of finite multiplicity $q^j(x)$.

2.2   Normal Form Matrices

We shall consider matrices

$$D : l^2(\mathcal{L}) \times X \to l^2(\mathcal{L})$$

which are covariant with respect to an $\mathcal{L}$-action

$$T : \mathcal{L} \times X \to X$$

and which satisfy the following conditions.
Exponential decay off the diagonal.

$$| D^b_a |_{C^0} \leq \beta e^{-\alpha |b-a|}. \tag{2.1}$$

Smoothness.   The components of D are piecewise smooth and satisfy

$$| D^b_a |_{C^k} \leq \beta e^{-\alpha |b-a|} \gamma^k \qquad \forall k \geq 1. \tag{2.2}$$

$D(x)$ is pure point with a (possibly generalized) eigenvector $q(x)$, corresponding eigenvalue $E(x)$ and block $\Omega(x)$ satisfying the following conditions.
Block dimensions and block extensions.   For all $x \in X$

$$\begin{cases} \Omega(x) \subset \{a : | a | \leq \lambda\} \\ \#\Omega(x) \leq \mu. \end{cases} \tag{2.3}$$

Block overlapping. For all $x \in X$

$$\# \bigcup_{\Omega^a(x) \cap \Omega(x) \neq 0} \Omega^a(x) \leq \mu. \qquad (2.4)$$

If the blocks do not overlap, then the matrix D is a block matrix with blocks of dimension $\leq \mu$. In general the blocks do overlap but we shall require that resonant blocks don't.

Resonant block separation. For all $x \in X$

$$| E^a(x) - E(x) | \leq \rho \Longrightarrow$$
$$\Omega^a(x) = \Omega(x) \text{ or } \underline{\text{dist}}(\Omega^a(x), \Omega(x)) \geq \nu \geq 1 \quad (2.5)$$

Partition. There is a locally finite partition $\mathcal{P}$ such that

$$D \text{ and } \Omega \text{ are smooth on the partition } \mathcal{P}. \qquad (2.6)$$

By a matrix D being smooth on a partition we mean that each D_a^b is smooth on, a neighborhood of the closure of, the pieces of

$$\vee \{T_c(\mathcal{P}) : | c + \frac{b+a}{2} | \leq | \frac{b-a}{2} | \}.$$

By a set Ω being smooth on a partition we mean that each the characteristic function $\chi_{\Omega^b}(a)$ is smooth on, a neighborhood of the closure of, the pieces of the same partition.

This smoothness condition is consistent with the covariance.

Remark. This concept of piecewise smoothness in relation to covariance is a little awkward. If A and B are two covariant matrices which are smooth on the partition, then AB will not be smooth on any partition unless A or B are truncated at some finite distance from the diagonal. If A or B are truncated at distance N, then AB is smooth on the refined partition

$$\mathcal{P}(N) =: \vee \{T_c(\mathcal{P}) : | c | \leq N \}.$$

Matrices like e^A or $(I+A)^{-1}$ will in general not be be piecewise smooth with the definition we have chosen.

Given a partition $\mathcal{P}$ we let $\mathcal{P}_\delta$ denote another type of refinement. Each piece Y of $\mathcal{P}_\delta$ is contained in a piece of $\mathcal{P}$ and has "diameter" less than δ, i.e. any two points x, y in Y can be joined by a curve *in* Y of length less than δ – if Y is convex then δ is the usual diameter of Y.

In the sequel we shall construct partitions like

$$\mathcal{P}(N_1)_{\delta_1}(N_2)_{\delta_2}(N_3)\ldots.$$

Definition. *We say that a covariant matrix D is on normal form*

$$\mathcal{NF}(\alpha, \beta, \gamma, \lambda, \mu, \nu, \rho)$$

if both D and D^* satisfy (2.1 − 6) with the same block Ω and partition $\mathcal{P}$. The eigenvalues will be complex conjugate, $E_{D^*} = \bar{E}_D$. The eigenvectors q_D and q_{D^*} of course not related unless D is symmetric, in which case they are equal, or Hermitian, in which case the are complex conjugate. We also use notations like

$$\mathcal{NF}(\alpha, \beta, \gamma, \lambda, \mu, \nu, \rho; q, E, \Omega, \mathcal{P})$$

when we want to stress these objects.

Remarks. 1. The parameters $\beta, \gamma, \lambda, \mu$ can be increased and the parameters α, ν, ρ can be decreased. For simplicity we shall assume that β, γ both are ≥ 1.

2. It follows from the definition that D is truncated at distance 2λ from the diagonal and that, for given a (or b), $D_a^b \neq 0$ for at most μ^2 many b's (or a's).

3. Using the generalized Young inequality [27] we get an estimate of D in the operator norm

$$| D |_{C^k} \leq \beta\left(\frac{e^\alpha + 1}{e^\alpha - 1}\right)^{\dim \mathcal{L}} \gamma^k \qquad \forall k \geq 0. \tag{2.7}$$

If D is a $m \times m$-dimensional matrix then we also have

$$| D |_{C^k} \leq m\beta\gamma^k \qquad \forall k \geq 0. \tag{2.8}$$

The estimate (2.8) is better than (2.7) if α is small but worse if m is large.

We have the corresponding notion for multi-level operators on $l^2(\mathcal{L}) \otimes \mathbb{C}^N$. We then have N eigenvectors q^j, each with corresponding eigenvalue E^j and block Ω^j. (2.1-2) should hold for the matrix elements $D_{a,i}^{b,j}$ and (2.3) should hold for the blocks. (2.4) and (2.5) takes the form: for all j and for all $x \in X$

$$\# \bigcup_{\Omega^{a,i}(x) \cap \Omega^j(x) \neq 0} \Omega^{a,i}(x) \leq \mu; \tag{2.4'}$$

for all i, j and for all $x \in X$

$$| E^{a,i}(x) - E^j(x) | \leq \rho \Longrightarrow$$
$$\Omega^{a,i}(x) = \Omega^j(x) \text{ or } \underline{dist}(\Omega^{a,i}(x), \Omega^j(x)) \geq \nu \geq 1. \tag{2.5'}$$

3   Block splitting

In this section we shall see that we can conjugate a normal form matrix to a true block diagonal matrix. The price to pay is in the smoothness and will be related to almost-multiplicity of the eigenvalues.

Proposition 1. *Let*

$$D \in \mathcal{NF}(\alpha, \beta, \gamma, \lambda, \mu, \nu, \rho; \Omega, \mathcal{P}).$$

Then there exists a subspace $\Lambda(x)$ *which is invariant under* $D(x)$ *such that all eigenvalues of* $\Lambda^* D\Lambda$ *are* ρ-close, *and which has the following properties:*

(i)   *for all* x

$$\Lambda(x) \subset \mathbb{C}^{\Omega(x)}; \tag{3.1}$$

(ii)

$$| \Lambda |_{C^k} \leq ((\text{const} \frac{\beta\mu^2}{\rho})^{2\mu} \gamma)^k \qquad \forall k \geq 0; \tag{3.2}$$

(iii) Λ *is smooth on the partition* $\mathcal{P}(3\lambda)_\delta$, *where*

$$\delta = (\text{const} \frac{\rho}{\beta\mu^2})^{2\mu} \frac{1}{\gamma}; \tag{3.3}$$

(iv) *the angle between* $\Lambda(x)$ *and the invariant subspace*

$$\sum_{\Lambda^b(x) \neq \Lambda(x)} \Lambda^b(x), \qquad \Lambda^b(x) = \tau_b \Lambda(T_b x),$$

is

$$\geq (\text{const} \frac{\rho}{\beta\mu^2})^{\mu^2}. \tag{3.4}$$

The constants only depend on the dimensions of $\mathcal{L}$ *and* X.

Proof. Let $\tilde{\Omega}(x) = \cup \Omega^a(x)$ where the union is taken over all a such that $\Omega^a(x) \cap \Omega(x) \neq \emptyset$. Then by $(2.3 - 4)$, for each x,

$$\begin{cases} \#\tilde{\Omega}(x) \leq \mu \\ \tilde{\Omega}(x) \subset \{a \in \mathcal{L} :| a |\leq 3\lambda\}. \end{cases}$$

Consider $D_{\tilde{\Omega}}$. This matrix is smooth on $\mathcal{P}(3\lambda)$ and by the remark (2.8) we have the estimate

$$| D_{\tilde{\Omega}} |_{C^k} \leq \mu\beta\gamma^k \qquad \forall k \geq 0.$$

Apply Lemma A.6 with $\rho = r$ to get a $D_{\tilde{\Omega}}(x)$-invariant decomposition

$$\mathbb{C}^{\tilde{\Omega}(x)} = \sum_{i=1}^{k} \tilde{\Lambda}_i(x)$$

smooth on $\mathcal{P}(3\lambda)_\delta$, with δ satisfying (3.3), and

$$| \tilde{\Lambda}_i |_{C^k} \leq ((\text{const} \frac{\beta\mu^2}{\rho})^{2\mu} \gamma)^k \qquad \forall k \geq 0.$$

$q(x)$ belongs to one of these spaces – $\tilde{\Lambda}_1(x)$ say. Let now x be fixed. If $\tilde{q}$ is an eigenvector of $D_{\tilde{\Omega}}(x)$ which lies in $\tilde{\Lambda}_1(x)$ then it follows from Lemma A.7 that either $\tilde{q}$ is an eigenvector of $D(x)$ or is perpendicular to $\mathbb{C}^{\Omega(x)}$. Hence, if $\tilde{q}_a \neq 0$ for some $a \in \Omega(x)$, then $\tilde{q}$ is an eigenvector of $D(x)$ and hence $= q^b(x)$ for some b. This implies that $| E(x) - E^b(x) | \leq \rho$ and by (2.5) that $\Omega^b(x) = \Omega(x)$. Hence, either $\tilde{q}$ is supported in $\Omega(x)$, in which case it is an eigenvector of $D(x)$, or it is supported in the complement of $\Omega(x)$. Therefore $\Lambda(x) = \mathbb{C}^{\Omega(x)} \cap \tilde{\Lambda}_1(x)$ is an invariant space for $D(x)$ which satisfies the same estimates as $\tilde{\Lambda}_1$, i.e. (3.2-3). This proves (i), (ii) and (iii).

The angle between Λ and Λ^b follows also from Lemma A.6 since each $\Lambda^b(x)$, which is not orthogonal to $\Lambda(x)$, is contained in $\mathbb{C}^{\tilde{\Omega}(x)}$ and therefore is a subspace of some $\tilde{\Lambda}_i$. $\square$

When D is Hermitian in Proposition 1, then the angle between the invariant spaces is $\frac{\pi}{2}$; the radius δ in (3.2) is

$$\text{const} \frac{\rho}{\beta\gamma\mu^2}.$$

In the multi-level case we get one subspace $\Lambda^j(x)$ for each $j = 1, \ldots, N$ with

$$\Lambda^j(x) \subset \mathbb{C}^{\Omega^j(x)},$$

and the matrix Q is defined by $Q(x) = (\ldots \Lambda^b(x) \ldots)$, where Λ^b is the subspace $(\Lambda^{b,1}, \ldots, \Lambda^{b,N})$. All the estimates are the same and the proof is also the same with obvious modifications.

4   Quadratic convergence

In this section we shall construct a conjugation which transforms a normal form plus a perturbation to a new normal form plus a smaller perturbation. The formulation will involve an auxiliary matrix W. The reason for this matrix is that we are not allowed to take inverses because they are not piecewise smooth, but we have to use approximate inverses. The role of W is to measure this approximation.

Lemma 2. *Let $D \in \mathcal{NF}(\alpha, \ldots, \rho; E, \mathcal{P})$ and let F be a covariant matrix, smooth on $\mathcal{P}$ and satisfying*

$$| F_a^b |_{C^k} \le \varepsilon e^{-\alpha|b-a|} \gamma^k \qquad \forall k \ge 0. \tag{4.1}$$

Then there exist covariant matrices K and G satisfying

$$\begin{cases} [D, K] = F - G \\ < q_{D^*}^a(x), G(x) q_D^b(x) > = 0 \quad if \ | E^a(x) - E^b(x) | > \rho, \end{cases}$$

with the following properties:

(i) *for all $k \ge 0$*

$$\begin{cases} | K_a^b |_{C^k} + | G_a^b |_{C^k} \le \varepsilon (\frac{1}{\beta\mu^2})(\text{const}\frac{\beta\mu^2}{\rho})^{3\mu^3} e^{6\lambda\alpha} e^{-\alpha|b-a|} (\gamma')^k \\ \gamma' = (\text{const}\frac{\beta\mu^2}{\rho})^{3\mu^3 + 4\mu^2} \gamma \end{cases} \tag{4.2}$$

(ii) *K and G are smooth on the partition $\mathcal{P}(3\lambda)_\delta(6\lambda)$ where*

$$\delta = (\text{const}\frac{\rho}{\beta\mu^2})^{2\mu} \frac{1}{\gamma}; \tag{4.3}$$

(iii) *if F is truncated at distance λ' from the diagonal, then K and G are truncated at distance $\lambda' + 6\lambda$.*

The constants only depend on the dimensions of $\mathcal{L}$ and X.

Proof. Let Λ be the invariant space of D defined in Proposition 1, with ρ replaced by $\frac{\rho}{3}$, and let

$$Q = (\ldots \Lambda^a \Lambda^b \ldots).$$

The estimate of Q follows from that of Λ, and Q is truncated at distance λ from the diagonal.

Let

$$\tilde{Q} = (\ldots \tilde{\Lambda}^a \tilde{\Lambda}^b \ldots)$$

be the same matrix as Q but obtained from D^*. The inverse $Q^{-1} = B\tilde{Q}^*$, where B is a block matrix with the blocks

$$[(\tilde{\Lambda}^a)^* \Lambda^a]^{-1}.$$

It follows that Q^{-1} is truncated at distance 2λ from the diagonal, is smooth on $\mathcal{P}(3\lambda)_\delta(\lambda)$ and satisfies

$$| Q^{-1} |_{C^k} \le (\text{const}\frac{\beta\mu^2}{\rho})^{2\mu^2} ((\text{const}\frac{\beta\mu^2}{\rho})^{4\mu^2} \gamma)^k \qquad \forall k \ge 0$$

– see Lemma A.6.

Then $\tilde{D} = Q^{-1}DQ$ is a block diagonal matrix with blocks – which we here denote by $\tilde{D}_a^a$ – corresponding to the subspaces Λ^a, which is smooth on $\mathcal{P}(3\lambda)_\delta(3\lambda)$ and satisfies

$$| \tilde{D}_a^b |_{C^k} \leq \beta\mu(\text{const}\frac{\beta\mu^2}{\rho})^{2\mu^2} e^{3\lambda\alpha} e^{-\alpha|b-a|}((\text{const}\frac{\beta\mu^2}{\rho})^{4\mu^2}\gamma)^k \qquad \forall k \geq 0.$$

Eigenvalues of $\tilde{D}_a^a$ are $\frac{\varrho}{3}$-close, so if $\tilde{D}_a^a$ and $\tilde{D}_b^b$ have two eigenvalues that differ by at least ρ then any two eigenvalues of $\tilde{D}_a^a$ and $\tilde{D}_b^b$ respectively differ by at least $\frac{\varrho}{3}$.

Let now $\tilde{F} = Q^{-1}FQ$. Then $\tilde{F}$ will be smooth on the same partition as $\tilde{D}$ and satisfy the same estimate but with the first factor β replaced by ε.

The equation becomes

$$\tilde{D}_a^a \tilde{K}_a^b - \tilde{K}_a^b \tilde{D}_b^b = \tilde{F}_a^b - \tilde{G}_a^b.$$

If any two eigenvalues of $\tilde{D}_a^a$ and $\tilde{D}_b^b$ respectively differ by $\leq \rho$, then the equation reduces to

$$\tilde{K}_a^b = 0 \quad \text{and} \quad \tilde{G}_a^b = \tilde{F}_a^b,$$

and if at least two eigenvalues differ by $> \rho$, then the equation becomes

$$\tilde{D}_a^a \tilde{K}_a^b - \tilde{K}_a^b \tilde{D}_b^b = \tilde{F}_a^b \quad \text{and} \quad \tilde{G}_a^b = 0.$$

The estimate of $\tilde{G}$ is clear so we only need to estimate $\tilde{K}$ from the second equation. Since we can triangularize each $\tilde{D}_a^a$ by a unitary transformation we get

$$| \tilde{K}_a^b |_{C^0} \leq \varepsilon[\frac{\beta\mu}{\rho}(\text{const}\frac{\beta\mu^2}{\rho})^{2\mu^2}]^\mu e^{3\lambda\alpha} e^{-\alpha|b-a|}.$$

Since the solution of the second equation is unique we get estimate of the derivatives by differentiating the equation.

We now let $K = Q\tilde{K}Q^{-1}$ and $G = Q\tilde{G}Q^{-1}$. They will satisfy the required estimates and be smooth on $\mathcal{P}(3\lambda)_\delta(6\lambda)$.

(iii) is clear by construction. $\square$

Proposition 3. *Let $D \in \mathcal{NF}(\alpha, \ldots, \rho; \Omega, \mathcal{P})$ and let F and W be covariant matrices, smooth on $\mathcal{P}$ and satisfying*

$$| F_a^b |_{C^k} + | W_a^b |_{C^k} \leq \varepsilon e^{-\alpha|b-a|}\gamma^k \qquad \forall k \geq 0.$$

Then there exists a constant C – depending only on the dimensions of X and $\mathcal{L}$ – such that if

$$\varepsilon \leq \xi(\frac{\rho - \rho'}{\beta}(\alpha - \alpha')^{\dim \mathcal{L}})^\mu, \qquad \xi =: (C\frac{\rho}{\beta\mu^2})^{3\mu^3} e^{-8\lambda\alpha}, \qquad (4.4)$$

then there exist

$$D' \in \mathcal{NF}(\alpha', \beta', \gamma', \lambda, \mu, \nu, \rho'; \Omega, \mathcal{P}')$$

and covariant matrices U, V, F' and W' such that

$$V(D+F)U = D' + F' \quad and \quad V(I+W)U = I + W'$$

with the following properties:

(i)

$$\begin{cases} \alpha' < \alpha, & \beta' = (1 + \frac{\varepsilon}{\xi})\beta, \\ \gamma' = (\frac{1}{C} \frac{\beta\mu^2}{\rho})^{3\mu^3 + 4\mu^2}\gamma, & \rho' < \rho; \end{cases} \tag{4.5}$$

(ii) *for all $k \geq 0$,*

$$| (U-I)_a^b |_{C^k} + | (V-I)_a^b |_{C^k} + | (D'-D)_a^b |_{C^k} \leq \frac{\varepsilon}{\xi} e^{-\alpha|b-a|}(\gamma')^k, \tag{4.6}$$

and $D' - D$ is truncated at distance ν from the diagonal;

(iii) *for all $k \geq 0$,*

$$\begin{cases} | (F')_a^b |_{C^k} + | (W')_a^b |_{C^k} \leq \varepsilon' e^{-\alpha'|b-a|}(\gamma')^k \\ \varepsilon' = \max[\frac{\varepsilon^2}{\xi^2(\alpha-\alpha')^{\dim L}}, \varepsilon e^{-(\nu-8\lambda)(\alpha-\alpha')}]; \end{cases} \tag{4.7}$$

(iv) *F' and W' are smooth on $\mathcal{P}' = \mathcal{P}(5\lambda)_\delta(6\lambda + 2\nu)$ where*

$$\delta = (C\frac{\rho}{\beta\mu^2})^{2\mu}\frac{1}{\gamma}. \tag{4.8}$$

Notice that the smallness assumption only depends on $\alpha, \beta, \lambda, \mu, \rho, \alpha - \alpha'$, $\rho - \rho'$ and that the size of the new perturbation F' depends on ν. Nothing depends on γ.

Proof. Let $\hat{F}$ and $\hat{W}$ be the truncations of F and W at distance $\nu - 8\lambda$ from the diagonal.

Define K and G so that

$$[K, D] + G = \hat{F} - \frac{1}{2}(\hat{W}D + D\hat{W}).$$

(This choice gives an Hermitian G and an anti-Hermitian K whenever D, F, W are Hermitian.) The right hand side is smooth on $\mathcal{P}(2\lambda)$, truncated at distance $\nu - 6\lambda$ from the diagonal and satisfies

$$| (RHS)_a^b |_{C^k} \leq const\varepsilon\beta\mu^2 e^{2\lambda\alpha}e^{-\alpha|b-a|}\gamma^k \qquad k \geq 0,$$

because D is truncated at distance 2λ and $D_c^d \neq 0$ for at most μ^2 many c's or d's.

Hence, by Lemma 2 we have

$$| K_a^b |_{C^k} + | G_a^b |_{C^k} \le \frac{\varepsilon}{\xi} e^{-\alpha|b-a|} (\gamma')^k$$

and K and G are smooth on the partition $\mathcal{P}(5\lambda)_\delta(6\lambda)$. Moreover, K and G are both truncated at distance ν.

If we define

$$U = I + K - \frac{1}{2}\hat{W}, \quad V = I - K - \frac{1}{2}\hat{W}$$

then U, V and G will satisfy the estimate (4.6) and

$$V(I + W)U - I = [(W - \hat{W}) + \mathbf{O}^2(W, \hat{W}, K)] + [\mathbf{O}^3(W, \hat{W}, K)]$$
$$=: W' = W_2' + W_3'$$

and

$$V(D + F)U - D - G = [(F - \hat{F}) + \mathbf{O}^2(F, \hat{W}, K)] + [\mathbf{O}^3(F, \hat{W}, K)]$$
$$=: F' = F_2' + F_3'.$$

Estimating this using Lemma A.8 gives that F_j' and W_j' satisfies (4.7). Observing that W' and F' are at most linear in the non-truncated matrices W, F we see that they are smooth on the partition

$$\mathcal{P}(5\lambda)_\delta(6\lambda + 2\nu).$$

It is clear that $D' = D + G$ satisfies (2.1-4) with $\alpha', \beta', \gamma', \lambda, \mu$ and (2.6) with $\mathcal{P}'$

Since the eigenvalues of D'_Ω and D_Ω differ by at most

$$4(\text{const}\frac{\beta}{\alpha^{\dim \mathcal{L}}})^{1-\frac{1}{\mu}} (\text{const}\frac{\varepsilon}{\xi \alpha^{\dim \mathcal{L}}})^{\frac{1}{\mu}} \le \frac{1}{4}(\rho - \rho')$$

(Lemma A.2), where we used (2.7) to estimate the norm of the matrices, it follows that

$$| (E')^a(x) - (E')(x) | \le \rho' \to | E^a(x) - E(x) | \le \rho.$$

Therefore D' verifies (2.5). □

We now iterate the construction in Proposition 3 a finite number of times in order to improve the result.

Corollary 4. *Let $D \in \mathcal{NF}(\alpha, \dots, \rho; \Omega, \mathcal{P})$ and let F and W be covariant matrices, smooth on $\mathcal{P}$ and satisfying*

$$| F_a^b |_{C^k} + | W_a^b |_{C^k} \le \varepsilon e^{-\alpha|b-a|} \gamma^k \qquad \forall k \ge 0.$$

Then there exists a constant C – depending only on the dimensions of X and $\mathcal{L}$ – such that if

$$\begin{cases} \varepsilon \leq \min[\xi^2(\alpha - \alpha')^{\dim \mathcal{L}}, \xi(\frac{\rho - \rho'}{\beta}(\alpha - \alpha')^{\dim \mathcal{L}})^\mu] \\ \xi =: (C\frac{\rho}{\beta\mu^2})^{3\mu^3}e^{-8\lambda\alpha}, \end{cases} \tag{4.9}$$

then, for any integer

$$1 \leq n \leq \frac{1}{2\log 2}[\log((\nu - 8\lambda)(\alpha - \alpha')) - \log\log(\frac{1}{\varepsilon})],$$

there exist

$$D' \in \mathcal{NF}(\alpha', \beta', \gamma', \lambda, \mu, \nu, \rho'; \Omega, \mathcal{P}')$$

and covariant matrices U, V, F' and W' such that

$$V(D + F)U = D' + F' \quad \text{and} \quad V(I + W)U = I + W'$$

with the following properties:

(i)

$$\begin{cases} \alpha' < \alpha, & \beta' = (1 + \frac{\varepsilon}{\xi})\beta \\ \gamma' = (\frac{1}{C}\frac{\beta\mu^2}{\rho})^{3\mu^3 + 4\mu^2}\gamma, & \rho' < \rho; \end{cases} \tag{4.10}$$

(ii) for all $k \geq 0$,

$$\mid (U - I)_a^b \mid_{C^k} + \mid (V - I)_a^b \mid_{C^k} + \mid (D' - D)_a^b \mid_{C^k} \leq \frac{\varepsilon}{\xi}e^{-\frac{\alpha}{2}|b-a|}(\gamma')^k, \tag{4.11}$$

and $D' - D$ is truncated at distance ν from the diagonal;
(iii) for all $k \geq 0$,

$$\begin{cases} \mid (F')_a^b \mid_{C^k} + \mid (W')_a^b \mid_{C^k} \leq \varepsilon'e^{-\alpha'|b-a|}(\gamma')^k \\ \varepsilon' = (\frac{\varepsilon}{\xi^2(\alpha - \alpha')^{\dim \mathcal{L}}})^{2^n}; \end{cases} \tag{4.12}$$

(iv) F' and W' are smooth on $\mathcal{P}' = \mathcal{R}^n(\mathcal{P})$, where

$$\begin{cases} \mathcal{R}(\mathcal{P}) = \mathcal{P}(5\lambda)_\delta(6\lambda + 2\nu) \\ \delta = (C\frac{\rho}{\beta\mu^2})^{2\mu}\frac{1}{\gamma}. \end{cases} \tag{4.13}$$

Proof. We can assume that $2\alpha' > \alpha$. Let $\alpha = \alpha_1 > \alpha_2 > \ldots$ and $\rho = \rho_1 > \rho_2 > \ldots$ with

$$\alpha_i - \alpha_{i+1} = 2^{-i}(\alpha - \alpha'), \qquad \rho_i - \rho_{i+1} = 2^{-i}(\rho - \rho'),$$

and let

$$\varepsilon_{i+1} = \max\left[\frac{\varepsilon_i^2}{\xi^2(\alpha_i - \alpha_{i+1})^{\dim \mathcal{L}}}, \varepsilon_i e^{-(\nu - 8\lambda)(\alpha_i - \alpha_{i+1})}\right]$$

$$= \frac{\varepsilon_i^2}{\xi^2(\alpha_i - \alpha_{i+1})^{\dim \mathcal{L}}} \qquad i \leq n.$$

If we apply n times Proposition 3 we get the statement (i-iv), but only with

$$\gamma' = \left(\frac{1}{C}\frac{\beta\mu^2}{\rho}\right)^{(3\mu^3 + 4\mu^2)n}\gamma.$$

In order to get better smoothness we must look more closely. Let us define

$$\begin{cases} V_j \ldots V_1(D + F)U_1 \ldots U_j = D_{j+1} + F_{j+1} \\ V_j \ldots V_1(I + W)U_1 \ldots U_j = I + W_{j+1}. \end{cases}$$

with

$$U_j = I + K_j - \frac{1}{2}\hat{W}_j, \quad V_j = I - K_j - \frac{1}{2}\hat{W}_j \quad D_{j+1} = D_j + \hat{G}_j$$

and

$$[K_j, D_j] + (D_{j+1} - D_j) = \hat{F}_j - \frac{1}{2}(\hat{W}_j D_j + D_j \hat{W}_j),$$

where ˆ denotes truncation at distance $\nu - 8\lambda$ from the diagonal and $D_1 = D$ and $F_1 = F$.

We now conjugate these equations to the left and right by Q^{-1} and Q. This conjugation will block diagonalize all the D_j's to $\tilde{D}_j$ and give the equations

$$[\tilde{K}_j, \tilde{D}_j] + \tilde{G}_j = Q^{-1}\hat{F}_j Q - \frac{1}{2}(Q^{-1}\hat{W}_j Q\tilde{D}_j + \tilde{D}_j Q^{-1}\hat{W}_j Q) = \tilde{R}HS.$$

These equations now split into blocks:

$$(\tilde{K}_j)_a^b = 0 \quad \text{and} \quad (\tilde{G}_j)_a^b = (\tilde{R}HS)_a^b$$

when any two eigenvalues of the blocks $\tilde{D}_a^a$ and $\tilde{D}_b^b$ differ by $\leq \rho$;

$$(\tilde{K}_j)_a^b(\tilde{D}_j)_b^b - (\tilde{D}_j)_a^a(\tilde{K}_j)_a^b = (\tilde{R}HS)_a^b \quad \text{and} \quad (\tilde{G}_j)_a^b = 0,$$

when at least two eigenvalues of the blocks $\tilde{D}_a^a$ and $\tilde{D}_b^b$ differ by $\geq \rho$. That the K_j's, D_j's and F_j's are γ'-smooth now follows by an easy induction. $\square$

When D, F and W are Hermitian these results can be improved. In Lemma 2, G is Hermitian and K is anti-Hermitian – this is an immediate consequence of the construction we have made; the estimate in (4.2) becomes

$$\begin{cases} |K_a^b|_{C^k} + |G_a^b|_{C^k} \leq \text{const}\,\varepsilon\frac{\beta\mu}{\rho}e^{4\lambda\alpha}e^{-\alpha|b-a|}(\gamma')^k \\ \gamma' = (\text{const}\frac{\beta\mu^2}{\rho})^{2\mu+1}\gamma; \end{cases}$$

the estimate for δ in (4.3) becomes $(\text{const} \frac{\rho}{\beta\gamma\mu^2})$.

In Proposition 3 and Corollary 4 D', F' and W' are Hermitian and $V = U^*$; the smallness assumptions (4.4) and (4.9) are given by

$$\varepsilon \leq \xi(\frac{\rho - \rho'}{\beta}(\alpha - \alpha')^{\dim \mathcal{L}}), \qquad \xi = (C\frac{\rho}{\beta\mu^2})e^{-6\lambda\alpha}$$

and

$$\varepsilon \leq \min[\xi^2(\alpha - \alpha')^{\dim \mathcal{L}}, \xi(\frac{\rho - \rho'}{\beta}(\alpha - \alpha')^{\dim \mathcal{L}})]$$

respectively; the estimate for γ' in (4.5) and (4.10) becomes

$$\gamma' = (\frac{\beta\mu^2}{C\rho})^{2\mu+1}\gamma;$$

the estimate of δ in (4.8) and (4.12) becomes $(C\frac{\rho}{\beta\gamma\mu^2})$.

In the multi-level case the proof goes through in the same way and all the results remain the same with obvious modifications.

5   Block clustering

The result in Corollary 4 depends on ν – it is better the larger ν is. In general we cannot increase ν, but it may happen that the Ω-blocks of

$$D \in \mathcal{NF}(\alpha, \beta, \gamma, \lambda, \mu, \nu, \rho; E, \Omega)$$

cluster into bigger blocks with a better separation property. Indeed it may happen that there are unions of blocks $\Omega' = \cup_i \Omega^{a_i}$ – the union depends on x – such that

$$| E^b(x) - E^c(x) | \leq \rho \Longrightarrow$$
$$(\Omega')^b(x) = (\Omega')^c(x) \text{ or } \underline{\text{dist}}((\Omega')^b(x), (\Omega')^c(x)) \geq \nu'. \quad (5.1)$$

Of course $\Omega'(x)$ is likely to be bigger so

$$\Omega'(x) \subset \{a :| a | \leq \lambda'\} \quad \text{and} \quad \#\Omega' \leq \mu'. \qquad (5.2)$$

Since D is on normal form and therefore satisfies (2.5) it is natural to expect that

$$\underline{\text{dist}}(\Omega^{a_i}(x), \Omega^{a_j}(x)) \geq \nu \qquad (5.3)$$

for all smaller blocks Ω^{a_i} building up Ω'.

Definition. *The Ω-blocks of $D \in \mathcal{NF}(\alpha, \beta, \gamma, \lambda, \mu, \nu, \rho; E, \Omega)$ are said to be*

$$\mathcal{C}(\lambda', \mu', \nu') - \text{clustering into } \Omega' - \text{blocks}$$

if (5.1-3) hold.

Then we get a better result.

Proposition 5. *Let $D \in \mathcal{NF}(\alpha, \dots, \rho; \Omega, \mathcal{P})$ be such that the Ω-blocks are $C(\lambda', \mu', \nu')$-clustering into Ω'-blocks. Let F and W be covariant matrices, both smooth on $\mathcal{P}$ and satisfying*

$$| F_a^b |_{C^k} + | W_a^b |_{C^k} \leq \varepsilon e^{-\alpha|b-a|}\gamma^k \qquad k \geq 0.$$

Then there exists a constant C – depending only on the dimensions of X and $\mathcal{L}$ – such that if

$$\begin{cases} \varepsilon \leq \min[\xi^4(\alpha - \alpha')^{2\dim \mathcal{L}}, \xi(\frac{\rho-\rho'}{\beta}(\alpha - \alpha')^{\dim \mathcal{L}})\mu'], \\ \xi =: (C\frac{\rho}{\beta\mu^2})^{3\mu^3} e^{-8\lambda\alpha}, \end{cases} \tag{5.4}$$

then there exist

$$D' \in \mathcal{NF}(\alpha', \beta', \gamma', \lambda', \mu', \nu', \rho'; \Omega', \mathcal{P}'),$$

truncated at distance ν' from the diagonal, and covariant matrices U, V, F' and W' such that

$$V(D + F)U = D' + F' \quad \text{and} \quad V(I + W)U = I + W'$$

with the following properties:

(i)

$$\begin{cases} \alpha' < \alpha, & \beta' = (1 + \sqrt{\varepsilon})\beta \\ \gamma' = (\frac{1}{C}\frac{\beta\mu^2}{\rho})^{7\mu^3}\gamma, & \rho' < \rho; \end{cases} \tag{5.5}$$

(ii) *for all $k \geq 0$*

$$| (U - I)_a^b |_{C^k} + | (V - I)_a^b |_{C^k} + | (D' - D)_a^b |_{C^k} \leq \sqrt{\varepsilon}\, e^{-\frac{\alpha}{2}|b-a|}(\gamma')^k \tag{5.6}$$

and $D' - D$ is truncated at distance ν' from the diagonal;

(iii) *for all $k \geq 0$,*

$$\begin{cases} | (F')_a^b |_{C^k} + | (W')_a^b |_{C^k} \leq \varepsilon' e^{-\alpha'|b-a|}(\gamma')^k \\ \varepsilon' = e^{-\frac{1}{2}\sqrt{(\nu'-8\lambda)(\alpha-\alpha')}}; \end{cases} \tag{5.7}$$

(iv) *F' and W' are smooth on*

$$\mathcal{P}' = \mathcal{R}^n(\mathcal{P}), \qquad \mathcal{R}(\mathcal{P}) = \mathcal{P}(5\lambda)_\delta(6\lambda + 2\nu'),$$

where

$$\begin{cases} n = \frac{1}{2\log 2}[\log((\nu' - 8\lambda)(\alpha - \alpha')) - \log\log(\frac{1}{\varepsilon})] \\ \delta = (C\frac{\rho}{\beta\mu^2})^{2\mu}\frac{1}{\gamma}. \end{cases} \tag{5.8}$$

Proof. The assumption implies that $D \in \mathcal{NF}(\alpha, \beta, \gamma, \lambda', \mu', \nu', \rho; \Omega', \mathcal{P})$. Applying n times Corollary 4 would give the result, but with a smallness condition (5.4) that depends on λ', μ'. But the occurrence of the block dimensions and block extensions enters through the block splitting and this only depends on λ, μ since $D \in \mathcal{NF}(\alpha, \beta, \gamma, \lambda, \mu, \nu, \rho; \Omega, \mathcal{P})$. The block dimension μ' enters however in the estimate of $|E' - E|$ which is reflected in the second smallness condition in (5.4).

Though Proposition 5 is not a corollary of Corollary 4, the proof is word by word the same as for Corollary 4. $\square$

5.1   Choice of parameters

How good must this clustering property be in order for us to apply Proposition 5 iteratively? We want sequences $\alpha_j, \beta_j, \gamma_j, \lambda_j, \mu_j, \nu_j, \rho_j, \varepsilon_j$ satisfying for all $j \geq 1$

$$\begin{cases} \beta_{j+1} = (1 + \sqrt{\varepsilon_j})\beta_j \\ \gamma_{j+1} = (\frac{1}{C}\frac{\beta_j \mu_j^2}{\rho_j})^{7\mu_j^3}\gamma_j \\ \varepsilon_{j+1} = e^{-\frac{1}{2}\sqrt{(\nu_{j+1} - 8\lambda_j)(\alpha_j - \alpha_{j+1})}} \end{cases} \tag{5.9}$$

$$\begin{cases} \varepsilon_j \leq \min[(C\frac{\rho_j}{\beta_j \mu_j^2})^{12\mu_j^3}e^{-32\lambda_j \alpha_j}(\alpha_j - \alpha_{j+1})^{2\dim\mathcal{L}}, \\ \qquad (\frac{\rho_j - \rho_{j+1}}{\beta_j}(\alpha_j - \alpha_{j+1})^{\dim\mathcal{L}})^{2\mu_{j+1}}] \\ \varepsilon_j \leq e^{-2j}\alpha_j^{2\dim\mathcal{L}}, \end{cases} \tag{5.10}$$

where C is the constant of Proposition 5.

If we assume that $\varepsilon_j \leq e^{-j}$, then $\beta_j \leq \text{const}\beta_1$ for all j, and it suffices that

$$\varepsilon_j \leq \min[(\frac{C\rho_j}{\beta_1 \mu_j^2})^{36\mu_j^3}, e^{-96\lambda_j \alpha_j}, (\alpha_j - \alpha_{j+1})^{6\dim\mathcal{L}\mu_{j+1}},$$

$$(\frac{\rho_j - \rho_{j+1}}{2\beta_1})^{4\mu_{j+1}}, e^{-4j}].$$

How shall we choose the sequences $\alpha_j, \lambda_j, \mu_j, \nu_j, \rho_j$ so that (5.10) is fulfilled for all $j \geq 1$. In order to simplify the discussion we assume a weak decay condition on the α_j's and the ρ_j's:

$$\alpha_j - \alpha_{j+1} \geq 4^{-j}\alpha_j \quad \text{and} \quad \rho_j - \rho_{j+1} \geq 4^{-j}\rho_j \qquad \forall j \geq 1.$$

These requirements leads to the conditions

$$\exp(-\frac{\sqrt{\nu_{j+1}\alpha_j}}{2^{j+6}\mu_{j+2}\dim\mathcal{L}}) \leq \alpha_{j+1} \leq \min[\frac{4^j - 1}{4^j}\alpha_j, \frac{\sqrt{\nu_{j+1}\alpha_j}}{2^{j+6}\lambda_{j+1}}] \tag{5.11}$$

and

$$exp(-\frac{\sqrt{\nu_{j+1}\alpha_j}}{2^{j+6}\mu_{j+2}}) \leq \frac{\rho_{j+1}}{\beta_1} \leq \frac{4^j-1}{4^j}\frac{\rho_j}{\beta_1} \qquad (5.12)$$

and

$$\sqrt{\nu_{j+1}\alpha_j} \geq 2^{j+6}\dim\mathcal{L}\begin{cases}\log(\frac{1}{\alpha_j})\log(\lambda_{j+1})\mu_{j+2}\\\log(\frac{\beta_1}{\rho_j})\mu_{j+2}\\\sqrt{\lambda_j\alpha_j}\\\log(\frac{2\beta_1\mu_{j+1}^2}{C})\mu_{j+1}^3.\end{cases} \qquad (5.13)$$

where the first two conditions make (5.11-12) possible.

Hence, if the $\mathcal{C}(\lambda_{j+1},\mu_{j+1},\nu_{j+1})$-clustering satisfies (5.13), then we can choose α_{j+1} and ρ_{j+1} according to (5.11-12), and (5.10) will hold for all $j \geq 1$ if it holds for $j = 1$. Notice that in the Hermitian case μ_{j+2} can be replaced by μ_{j+1} in (5.11-13).

Proposition 6. *Let $D_1 \in \mathcal{NF}(\alpha_1,\ldots,\rho_1;E_1,\Omega_1,\mathcal{P}_1)$ and let W_1 and F_1 be covariant matrices, smooth on the $\mathcal{P}_1$, with $W_1 = 0$ and*

$$\mid (F_1)_a^b \mid_{C^k} \leq \varepsilon_1 e^{-\alpha_1|b-a|}\gamma_1^k \qquad \forall k \geq 0.$$

Then there exists a constant C – depending only on the dimensions of X and $\mathcal{L}$ – such that if the sequences $\alpha_j,\beta_j,\gamma_j,\lambda_j,\mu_j,\nu_j,\rho_j,\varepsilon_j$ satisfy (5.9-10) then the following hold.

(I) If the Ω_1-blocks are $\mathcal{C}(\lambda_2,\mu_2,\nu_2)$-clustering into Ω_2-blocks, then there is an $J \geq 2$ such that the statement $(\mathcal{S}_J)$ holds: for all $1 \leq j < J$ there exist

$$D_{j+1} \in \mathcal{NF}(\alpha_{j+1},\ldots,\rho_{j+1};E_{j+1},\Omega_{j+1},\mathcal{P}_{j+1}),$$

truncated at distance ν_{j+1} from the diagonal, and matrices U_j,V_j,

$$V_j(D_j+F_j)U_j = D_{j+1}+F_{j+1}, \qquad V_j(I+W_j)U_j = I+W_{j+1}$$

that satisfy $(5.6-8)_j$.

(II) If the Ω_J-blocks are $\mathcal{C}(\lambda_{J+1},\mu_{J+1},\nu_{J+1})$-clustering into Ω_{J+1}-blocks, then also $(\mathcal{S}_{J+1})$ holds.

(III) If $(\mathcal{S}_\infty)$ holds, then there exists a matrix U such that

$$U(x)^{-1}(D(x)+F(x))U(x) = D_\infty(x), \quad \forall x \in X,$$

and $D_\infty(x)$ is a norm limit of $D_j(x)$. If D and F are Hermitian, then U is unitary.

(IV) Let η_j be the measure of the set of all $x \in X$ such that

$$\#\{\mid a \mid \leq \nu_{j+1}+2\lambda_j :\mid E_j^a(x)-E_j(x)\mid \leq \rho_j\} > 1.$$

If $\sum \eta_j < \infty$, then $D_\infty(x)$ is pure point for a.e. x.

Proof. (I-III) are direct consequences of Proposition 5. We get U and U^{-1} as the norm limit of the compositions

$$U_1 U_2 \ldots U_j \quad \text{and} \quad V_j \ldots V_2 V_1.$$

If $\sum \eta_j < \infty$, it follows that for a.e x

$$\#\{|\,a\,| \leq \nu_{j+1} + 2\lambda_j : |\,E_j^a(x) - E_j(x)\,| \leq \rho_j\} = 1, \quad \forall j \geq j(x).$$

This implies that the blocks stop increasing, i.e.

$$\Omega_j(x) = \Omega_{j(x)}(x), \quad \forall j \geq j(x),$$

for a.e. $x \in X$. This proves (IV). □

In the multi-level case the proofs goes through in the same way and all the results remain the same with obvious modifications.

We can describe this perturbative process with a diagram.

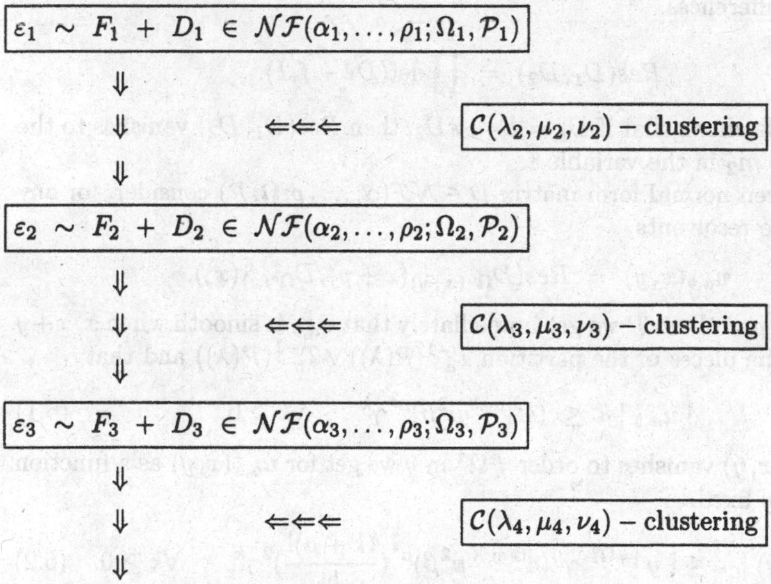

$$\boxed{\varepsilon_1 \sim F_1 + D_1 \in \mathcal{NF}(\alpha_1, \ldots, \rho_1; \Omega_1, \mathcal{P}_1)}$$

$$\Downarrow$$
$$\Downarrow \qquad \Leftarrow\Leftarrow\Leftarrow \qquad \boxed{\mathcal{C}(\lambda_2, \mu_2, \nu_2) - \text{clustering}}$$
$$\Downarrow$$

$$\boxed{\varepsilon_2 \sim F_2 + D_2 \in \mathcal{NF}(\alpha_2, \ldots, \rho_2; \Omega_2, \mathcal{P}_2)}$$

$$\Downarrow$$
$$\Downarrow \qquad \Leftarrow\Leftarrow\Leftarrow \qquad \boxed{\mathcal{C}(\lambda_3, \mu_3, \nu_3) - \text{clustering}}$$
$$\Downarrow$$

$$\boxed{\varepsilon_3 \sim F_3 + D_3 \in \mathcal{NF}(\alpha_3, \ldots, \rho_3; \Omega_3, \mathcal{P}_3)}$$

$$\Downarrow$$
$$\Downarrow \qquad \Leftarrow\Leftarrow\Leftarrow \qquad \boxed{\mathcal{C}(\lambda_4, \mu_4, \nu_4) - \text{clustering}}$$
$$\Downarrow$$

Figure 1

6   Transversality of resultants

In this section we shall derive some estimates of the resultants of the matrices D_j constructed in Proposition 6, which will prove essential in showing the clustering property of blocks.

The resultant $Res(D_1, D_2)$ of two square matrices D_1 and D_2 of dimension m_1 and m_2, respectively, is the product of the differences of the eigenvalues, i.e.

$$\prod(e_i - f_j),$$

where e_i and f_j are the eigenvalues of D_1 and D_2 respectively. All the eigenvalues are counted, so this is always a product of $m_1 m_2$ many factors. The resultant measures the difference between the spectra of the two matrices. For example

$$|e_1 - f_1| \leq \varepsilon \Longrightarrow |Res(D_1, D_2)| \leq \varepsilon(|D_1| + |D_2|)^{m_1 m_2 - 1}.$$

The resultant $Res(D_1, D_2) = \det \chi_{D_2}(D_1)$, where $\chi_{D_2}(t) = \det(D_2 - tI)$ is the characteristic polynomial of the matrix D_2. $\chi_{D_2}(t)$ is a sum of $\leq m_2!$ many monomials of degree m_2 in the matrix elements of $D_2 - tI$. Introducing D_1 for t we get an $m_1 \times m_1$-matrix whose elements are $\leq m_2!(m_1)^{m_2-1}$ many monomials of degree m_2 in the matrix elements of these two matrices and in their differences. Taking the determinant gives this number to the power m_1, times $m_1!$. This shows that $Res(D_1, D_2)$ is a sum of less than $(m_1 m_2)^{m_1 m_2}$ many monomials of degree less than $m_1 m_2$ in the components of D_1 and D_2 and their differences.

Moreover

$$Res(D_1, D_2) = \prod \det(D_1 - f_j I).$$

This formula shows that if $D_1 = D_2 + \varepsilon \hat{D}_2$, then $Res(D_1, D_2)$ vanishes to the order $m_1 = m_2$ in the variable ε.

For a given normal form matrix $D \in \mathcal{NF}(\alpha, \ldots, \rho; \Omega, \mathcal{P})$ consider, for any $a, b \in \mathcal{L}$, the resolvents

$$u_{a,b}(x, y) = Res(D_{\Omega^a(x+y)}(x+y), D_{\Omega^b(x)}(x)).$$

Since D is smooth on $\mathcal{P}$ we get immediately that $u_{a,b}$ is smooth when $x, x+y$ vary over the pieces of the partition $T_a^{-1}(\mathcal{P}(\lambda)) \vee T_b^{-1}(\mathcal{P}(\lambda))$ and that

$$|u_{a,b}|_{C^k} \leq (4^{\dim X} \mu^2 \beta)^{\mu^2} \gamma^k \qquad \forall k \geq 0. \qquad (6.1)$$

Since $u_{a,a}(x, y)$ vanishes to order $\#\Omega^a$ in y we get for $u_{a,a}(x, y)$ as a function of x, with y fixed,

$$|u_{a,a}(\,, y)|_{C^k} \leq |y|^{\#\Omega^a} \gamma^\mu (4^{\dim X} \mu^2 \beta)^{\mu^2} \left(\frac{(k+\mu)!}{k!}\right)^2 \gamma^k \qquad \forall k \geq 0. \qquad (6.2)$$

Consider now the conditions: for all $x, y \in X$ and for all i

$$\max_{0 \leq k \leq J} \left| \frac{1}{(k!)^2 \gamma^k} \partial_{y_i}^k u_{a,b}(x, y) \right| \geq \sigma, \qquad (6.3)$$

$$\max_{0 \leq k \leq J} \left| \frac{1}{(k!)^2 \gamma^k} \partial_{x_i}^k u_{a,b}(x, y) \right| \geq \sigma |y|^{\#(\Omega^a(x+y) \cap \Omega^b(x))}, \qquad (6.4)$$

where $J = \#\Omega^a(x+y) \times \#\Omega^b(x)s$ and s is a fixed parameter.

Definition. *We say that $D \in \mathcal{NF}(\alpha, \beta, \gamma, \lambda, \mu, \nu, \rho; \Omega)$ is (σ, s)-transversal on the Ω-blocks, denoted*

$$D \in \mathcal{NF}(\alpha, \beta, \gamma, \lambda, \mu, \nu, \rho; \Omega)\&\mathcal{T}(\sigma, s),$$

if the functions $u_{a,b}$ satisfy (6.3-4) for all x, y such that $\Omega^a(x + y)$ and $\Omega^b(x)$ are disjoint or equal.

Lemma 7. *Assume that*

$$D \in \mathcal{NF}(\alpha, \beta, \gamma, \lambda, \mu, \nu, \rho; \Omega, \mathcal{P})\&\mathcal{T}(\sigma, s)$$

is truncated at distance ν from the diagonal, and assume also that the Ω-blocks are $\mathcal{C}(\lambda', \mu', \nu')$-clustering into Ω'-blocks. Let

$$D' \in \mathcal{NF}(\alpha', \beta', \gamma', \lambda', \mu', \nu', \rho'; \Omega', \mathcal{P}')$$

be a perturbation of D

$$\mid (D' - D)_a^b \mid_{C^k} \leq \sqrt{\varepsilon} e^{-\frac{\alpha}{2}|b-a|}(\gamma')^k \qquad \forall k \geq 0.$$

If

$$\varepsilon \leq C^{(\mu')^4}(\frac{\gamma}{\gamma'})^{2s(\mu')^2}(\frac{\sigma}{\beta})^{4s(\mu')^4} \tag{6.5}$$

then

$$D' \in \mathcal{NF}(\alpha', \beta', \gamma', \lambda', \mu', \nu', \rho'; \Omega', \mathcal{P}')\&\mathcal{T}(\sigma', s)$$

for any

$$\sigma' \leq C^{(\mu')^4}(\frac{\gamma}{\gamma'})^{s(\mu')^2}(\frac{\sigma}{\beta})^{s(\mu')^4}. \tag{6.6}$$

The constant C depends only on the dimensions of $\mathcal{L}$ and X, and on s.

Proof. Consider first the case (6.3). Let

$$\tilde{u}_{c,d}(x, y) = Res(D_{(\Omega')^c(x+y)}(x+y), D_{(\Omega')^d(x)}(x))$$

where

$$\Omega'^c = \cup \Omega^a, \qquad (\Omega')^d = \cup \Omega^b,$$

are unions of less than μ' many Ω^a's which are separated by a distance at least ν. Since D is truncated at distance ν from the diagonal we get that

$$\tilde{u}_{c,d}(x, y) = \prod_{a,b} u_{a,b}.$$

Then by Lemma B2

$$\max_{0 \leq k \leq s(\mu')^2} \mid \frac{1}{(k!)^2(\gamma')^k}\partial_{y_i}^k \tilde{u}_{c,d}(x, y) \mid \geq (\frac{1}{2})^{8s^2(\mu')^4}(\frac{\gamma}{\gamma'})^{s(\mu')^2}(\frac{\sigma}{\beta})^{s(\mu')^4}. \tag{6.3}$$

Let now

$$u'_{c,d}(x,y) = Res(D'_{(\Omega')^c(x+y)}(x+y), D'_{(\Omega')^d(x)}(x)).$$

Then

$$| u'_{c,d}(x,) - \tilde{u}_{c,d}(x,) |_{C^k} \le \sqrt{\varepsilon}(4^{\dim X}(\mu')^2)^{(\mu')^2+1}(\beta+\sqrt{\varepsilon})^{(\mu')^2}(\gamma')^k \quad \forall k \ge 0$$

which gives the result.

The case (6.4), when the blocks are disjoint, is proven in the same way. In order to see (6.4) when the blocks are equal we consider $v_{a,b}(x,y) = u_{a,b}(x,y)$ when $\Omega^a(x+y) \cap \Omega^b(x) = \emptyset$ and

$$v_{a,b}(x,y) = u_{a,b}(x,y)\frac{1}{| y |^{\#\Omega^a(x+y)}},$$

when $\Omega^a(x+y) = \Omega^b(x)$.

We define $\tilde{v}_{c,d}$ and $v'_{c,d}$ in the same way as $\tilde{u}_{c,d}$ and $u'_{c,d}$ using v instead of u. We now apply the same argument to the v:s as before to the u's.   □

6.1   Choice of parameters

We want sequences $\alpha_j, \beta_j, \gamma_j, \lambda_j, \mu_j, \nu_j, \rho_j, \sigma_j, \varepsilon_j$ satisfying for all $j \ge 1$

$$\begin{cases} \beta_{j+1} = (1 + \sqrt{\varepsilon_j})\beta_j \\ \gamma_{j+1} = (\frac{1}{C}\frac{\beta_j\mu_j^2}{\rho_j})^{7\mu_j^3}\gamma_j \\ \varepsilon_{j+1} = e^{-\sqrt{\nu_{j+1}-8\lambda_j)(\alpha_j-\alpha_{j+1})}} \\ \sigma_{j+1} = C\mu_{j+1}^4(C\frac{\rho_j}{\beta_j\mu_j^2})^{14s\mu_j^3\mu_{j+1}^2}(\frac{\sigma_j}{\beta_j})^{s\mu_{j+1}^4}, \end{cases} \tag{6.7}$$

and, besides (6.8)=(5.10),

$$\varepsilon_j \le C\mu_{j+1}^4(\frac{\gamma_j}{\gamma_{j+1}})^{4s\mu_{j+1}^2}(\frac{\sigma_j}{\beta_j})^{2s\mu_{j+1}^4}, \tag{6.9}$$

where C is the smallest of the constant in Proposition 5 and Lemma 7.

How shall we choose the sequences $\lambda_j, \mu_j, \nu_j, \alpha_j, \rho_j$ so that the two conditions (6.8-9) are fulfilled? If we assume that

$$\mu_{j+1} \ge 4s\mu_j \qquad \forall j \ge 1$$

it follows that

$$\sigma_{j+1} \le (\frac{C\sigma_1\rho_j}{\beta_1})^{(\mu_1\dots\mu_{j+1})^5}.$$

Then we are lead to (6.10)=(5.11),

$$exp(-\frac{\sqrt{\nu_{j+1}\alpha_j}}{2^{j+6}(\mu_1\dots\mu_{j+2})^5}) \le \frac{\rho_{j+1}}{\beta_1} \le \frac{4^j-1}{4^j}\frac{\rho_j}{\beta_1} \tag{6.11}$$

and

$$\sqrt{\nu_{j+1}\alpha_j} \geq (2s)^{j+6}(\dim \mathcal{L}) \begin{cases} \log(\frac{1}{\alpha_j}) \log(\lambda_{j+1})\mu_{j+2} \\ \log(\frac{1}{\rho_j})(\mu_1 \ldots \mu_{j+2})^5 \\ \sqrt{\lambda_j \alpha_j} \\ \log(\frac{\beta_1 \mu_{j+1}^2}{C\sigma_1})(\mu_1 \ldots \mu_{j+2})^5. \end{cases} \tag{6.12}$$

The clustering required by (6.12) is much stronger than (5.13) but it will be fulfilled in the applications we have in mind.

Proposition 8. *Let $D_1 \in \mathcal{NF}(\alpha_1, \ldots, \rho_1; E_1, \Omega_1, \mathcal{P}_1)\&\mathcal{T}(\sigma_1, s)$ and assume that the D_1 is truncated at distance ν_1 from the diagonal. Let W_1 and F_1 be covariant matrices, smooth on $\mathcal{P}_1$ with $W_1 = 0$ and*

$$\mid (F_1)_a^b \mid_{C^k} \leq \varepsilon \gamma_1^k \qquad \forall k \geq 0.$$

Then there exists a constant C – depending only on the dimensions of X and $\mathcal{L}$ and on s – such that if the sequences $\alpha_j, \beta_j, \gamma_j, \lambda_j, \mu_j, \nu_j, \rho_j, \sigma_j, \varepsilon_j$ satisfy (6.7-9) then the following hold.

(I) If the Ω_1-blocks are $C(\lambda_2, \mu_2, \nu_2)$-clustering into Ω_2-blocks, then there is an $J \geq 2$ such that the statement (S'_J) holds: for all $1 \leq j < J$ there exist

$$D_{j+1} \in \mathcal{NF}(\alpha_{j+1}, \ldots, \rho_{j+1}; E_{j+1}, \Omega_{j+1}, \mathcal{P}_{j+1})\&\mathcal{T}(\sigma_{j+1}, s)$$

and matrices U_j, V_j,

$$V_j(D_j + F_j)U_j = D_{j+1} + F_{j+1}, \qquad V_j(I + W_j)U_j = I + W_{j+1}$$

that satisfy $(5.6 - 8)_j$.

(II) If the Ω_J-blocks are $C(\lambda_{J+1}, \mu_{J+1}, \nu_{J+1})$-clustering into Ω_{J+1}-blocks, then also (S'_{J+1}) holds.

(III) If (S'_∞) holds, then there exists a transformation U such that

$$U(x)^{-1}(D(x) + F(x))U(x) = D_\infty(x), \quad \forall x \in X,$$

and $D_\infty(x)$ is a norm limit of $D_j(x)$. If D and F are Hermitian, then U is unitary.

(IV) Let η_j be the measure of all $x \in X$ such that

$$\#\{\mid a \mid \leq \nu_{j+1} + 2\lambda_j : \mid E_j^a(x) - E_j(x) \mid \leq \rho_j\} > 1.$$

If $\sum \eta_j < \infty$, then $D_\infty(x)$ is pure point for a.e. x.

Proof. Immediate consequence of Proposition 6 and Lemma 7. □

In the multi-level case the transversality condition applies to the resultants

$$u_{(a,i),(b,j)}(x,y) = Res(D_{\Omega^{a,i}(x+y)}(x+y), D_{\Omega^{b,j}(x)}(x)).$$

The proofs goes through in the same way and all the results remain the same with obvious modifications.

We can describe this perturbative process with the diagram in Figure 2. In the next section we shall consider only matrices over $\mathbb{T}$ with a Diophantine quasi-periodic group action, and then we shall see that transversality on blocks implies clustering of blocks and the diagram will then be closed.

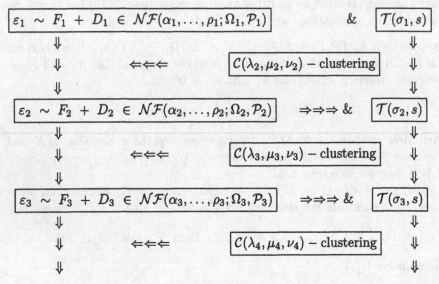

Figure 2

7   A Perturbation theorem

In this section we shall restrict the discussion to the circle $X = \mathbb{T}$ and the $\mathcal{L}$-action

$$\begin{cases} T : \mathcal{L} \times \mathbb{T} \to \mathbb{T} \\ (a, x) \mapsto (x + <a, \omega>) \end{cases} \tag{7.1}$$

where ω is a Diophantine vector

$$\| <a, \omega> \| \geq \frac{\kappa}{|a|^\tau} \qquad \forall a \in \mathcal{L} \setminus \{0\} \tag{7.2}$$

$- \tau > \dim \mathcal{L}$ and $\kappa > 0$ are fixed numbers. The numbers τ, κ will appear in the estimates but since they will be fixed throughout the iteration we shall not make them explicit in our notations.

We shall use the Diophantine property in Lemma 9 and also in Lemma 11 where we shall make a particular choice of the partitions $\mathcal{P}_j$. In Lemma 10 we make use of the first transversality condition (6.3), and in the proof of Theorem 12 we also use the second transversality condition (6.4).

Definition. *We say that the eigenvalues of $D \in \mathcal{NF}(\alpha, \ldots, \rho; E)\&\mathcal{T}(\sigma, s)$ have almost-multiplicity $\leq \frac{\mu'}{\mu}$ if, for all x and for all $t \leq \rho$, the inequality*

$$| E(x+y) - E(x) | > t(\frac{\mu'}{\mu})$$

is fulfilled outside at most $\frac{\mu'}{\mu}$ many intervals of length less than

$$2(4\beta\mu')^{\frac{1}{s}}(\frac{t}{\sigma})^{\frac{1}{s\mu^2}}.$$

Lemma 9. *Let the eigenvalues of*

$$D \in \mathcal{NF}(\alpha, \beta, \gamma, \lambda, \mu, \nu, \rho; E, \Omega, P)\&\mathcal{T}(\sigma, s)$$

have almost-multiplicity $\leq \frac{\mu'}{\mu}$ and assume

$$\rho \leq (\frac{\kappa}{4\beta\mu'}\frac{1}{\lambda})^{2\tau s\mu^2}\sigma. \tag{7.3}$$

Then

$$D \in \mathcal{NF}(\alpha, \beta, \gamma, \lambda, \mu, \nu, \rho; E, \Omega, P)\&\mathcal{C}(\lambda', \mu', \nu')$$

for any

$$\nu' = \frac{\mu}{2\mu'}(\frac{\kappa}{4\beta\mu'})^{\frac{1}{\tau}}(\frac{\sigma}{\rho})^{\frac{1}{\tau s\mu^2}}, \qquad \lambda' = \frac{\mu'}{\mu}(\nu' + 2\lambda). \tag{7.4}$$

Proof. For each x,

$$| E(x+y) - E(x) | > \rho(\frac{\mu'}{\mu})$$

is fulfilled outside at most $\frac{\mu'}{\mu}$ many intervals of length less than

$$L = 2(4\beta\mu')^{\frac{1}{s}}(\frac{\rho}{\sigma})^{\frac{1}{s\mu^2}}.$$

If $\| < a, \omega > \| \leq L$ it follows that

$$|a| \geq (\frac{\kappa}{L})^{\frac{1}{\tau}} = S.$$

This means that on distances of size less than S there are at most $\frac{\mu'}{\mu}$ many ρ-almost resonant eigenvalues and, hence, at most $\frac{\mu'}{\mu}$ many resonant blocks. Each resonant block extends over 2λ sites so there must be a gap of size at least $\frac{\mu}{\mu'}(S - 2\lambda\frac{\mu'}{\mu}) \geq \frac{\mu}{2\mu'}S = \nu'$ between the resonant blocks. And unless there is a gap of size ν' between the resonant blocks, they will extend over distances which are at most $\lambda' = \frac{\mu'}{\mu}(\nu' + 2\lambda)$. This proves the lemma.   □

We must now take into account the partition $\mathcal{P}$ of T. We shall for this discussion fix an integer p and define $|\mathcal{P}|$ to be the minimum of the length of the pieces, with the exception of the $p-1$ smallest pieces. Hence, all pieces of $\mathcal{P}$ with the exception of at most $p-1$ pieces have a length larger than this number.

Lemma 10. *Let*

$$D \in \mathcal{NF}(\alpha,\beta,\gamma,\lambda,\mu,\nu,\rho;E,\Omega,\mathcal{P})\&\mathcal{T}(\sigma,s).$$

(i) Then the eigenvalues of D have almost-multiplicity $\leq \frac{\mu'}{\mu}$ with

$$\frac{\mu'}{\mu} = 2^{s\mu^2}16\frac{\gamma}{\sigma}(4\beta\mu^2)^{\mu^2}(s\mu^2)^2(p+|\mathcal{P}(\lambda)|^{-1}).$$

(ii) Assume that the Ω-blocks are $\mathcal{C}(\lambda',\mu',\nu')$-clustering into Ω'-blocks, and let

$$D' \in \mathcal{NF}(\alpha',\beta',\gamma',\lambda',\mu',\nu',\rho';E',\Omega',\mathcal{P}')\&\mathcal{T}(\sigma',s)$$

be a perturbation of D,

$$|(D-D')^b_a|_{\mathcal{C}^0} \leq \sqrt{\varepsilon}e^{-\alpha'|b-a|}.$$

If

$$\varepsilon \leq \min[(\frac{1}{2s\beta'\mu'})^{8s\mu^2(\mu')^3}(\frac{\sigma'}{\gamma'})^{4s\mu^2\mu'},|\mathcal{P}'(\lambda')|^{4s\mu^2\mu'}] \qquad (7.5)$$

and

$$\rho' \leq \varepsilon^{\frac{1}{\mu'}}, \qquad (7.6)$$

then the eigenvalues of D' have almost-multiplicity $\leq \frac{\mu''}{\mu'}$ with

$$\mu'' = p2^{2s(\mu')^2}. \qquad (7.7)$$

Proof. Let $u(x,y)$ be the resultant of $D_{\Omega(x+y)}(x+y)$ and $D_{\Omega(x)}(x)$. $u(x,y)$ is smooth for $y \in I$, $x+I \in \mathcal{P}(\lambda)$, and satisfies

$$|u(x,\ )|_{\mathcal{C}^k} \leq \tilde{\beta}(\gamma)^k = (4\beta\mu^2)^{\mu^2}\gamma^k \qquad \forall k \geq 0.$$

The eigenvalues of D are bounded by $\beta\mu$, so if

$$|E(x+y)-E(x)| \leq t(\frac{\mu'}{\mu}), \qquad t \leq \rho,$$

it follows that

$$|u(x,y)| \leq t(\frac{\mu'}{\mu})(2\beta\mu)^{\mu^2}.$$

We can now apply Lemma B1 to each interval I. It follows that, for each x and each $t \leq \rho$

$$| E(x + y) - E(x) | > t(\frac{\mu'}{\mu})$$

is fulfilled outside at most $\frac{\mu'}{\mu}$ many intervals of length less than

$$L_t = 2(4\beta\mu')^{\frac{1}{s}}(\frac{t}{\sigma})^{\frac{1}{s\mu^2}},$$

i.e. the eigenvalues of D have almost-multiplicity $\leq \frac{\mu'}{\mu}$.

In order to prove (ii) notice that the eigenvalues of $D'_{\Omega'}$ and $D_{\Omega'}$ differ by at most $2\beta'\mu'\varepsilon^{\frac{1}{2\mu'}}$. Using (i) it follows that

$$| E'(x + y) - E'(x) | > \frac{t}{2}(\frac{\mu'}{\mu})$$

is fulfilled outside at most $\frac{\mu'}{\mu}$ many intervals of length less than L_t if we let

$$t = 4\beta'\mu\varepsilon^{\frac{1}{2\mu'}}.$$

By the second part of (7.5), each such interval is cut by the partition $\mathcal{P}'(\lambda')$ into at most p subintervals I.

By (7.6), $\rho'(\frac{\mu''}{\mu'}) \leq \frac{t}{2}(\frac{\mu'}{\mu})$ so we can restrict to one of these intervals I. Let $u'(x, y)$ be the resultant of $D'_{\Omega'(x+y)}(x + y)$ and $D'_{\Omega'(x)}(x)$. $u'(x, y)$ is smooth for $y \in I$ and satisfies

$$| u'(x,) |_{C^k} \leq \tilde{\beta}'(\gamma')^k = (4\beta'(\mu')^2)^{(\mu')^2}(\gamma')^k \qquad \forall k \geq 0.$$

The eigenvalues of D' are bounded by $\beta'\mu'$, so if

$$| E'(x + y) - E'(x) | \leq t'(\frac{\mu''}{\mu'}), \qquad t' \leq \rho',$$

it follows that

$$| u'(x, y) | \leq t'(\frac{\mu''}{\mu'})(2\beta'\mu')^{(\mu')^2}.$$

By Lemma B1 this defines a union of at most

$$2^{s(\mu')^2}(\frac{\gamma'}{\sigma'}8\tilde{\beta}'(s(\mu')^2 + 1)^2 | I | + 1)$$

many intervals, each of which has length at most

$$2(4\beta'\mu'')^{\frac{1}{s}}(\frac{t'}{\sigma})^{\frac{1}{s(\mu')^2}}.$$

By the first condition in (7.5) the number of intervals is less than

$$\frac{1}{\mu'}2^{2s(\mu')^2}$$

$\square$

7.1   Choice of parameters

We want sequences $\alpha_j, \beta_j, \gamma_j, \lambda_j, \mu_j, \nu_j, \rho_j, \sigma_j, \varepsilon_j$ satisfying for all $j \geq 1$

$$
\begin{cases}
\beta_{j+1} = (1 + \sqrt{\varepsilon_j})\beta_j \\
\gamma_{j+1} = (\frac{1}{C}\frac{\beta_j \mu_j^2}{\rho_j})^{7\mu_j^3}\gamma_j \\
\varepsilon_{j+1} = e^{-\frac{1}{2}\sqrt{(\nu_{j+1}-8\lambda_j)(\alpha_j-\alpha_{j+1})}} \\
\sigma_{j+1} = C^{\mu_{j+1}^4}(C\frac{\rho_j}{\beta_j \mu_j^2})^{14s\mu_j^3 \mu_{j+1}^2}(\frac{\sigma_j}{\beta_j})^{s\mu_{j+1}^4} \\
\lambda_{j+1} = (\frac{\mu_{j+1}}{\mu_j})(\nu_{j+1} + 2\lambda_j) \\
\mu_{j+1} = \tilde{p}2^{2s\mu_j^2}, \qquad \tilde{p} = 16\frac{\gamma_1}{\sigma_1}(4\beta_1\mu_1^2)^{\mu_1^2}(s\mu_1^2)^2(p+\mid \mathcal{P}_1(\lambda_1)\mid^{-1}) \\
\nu_{j+1} = \frac{\mu_j}{2\mu_{j+1}}(\frac{\kappa}{8\beta_j \mu_{j+1}})^{\frac{1}{\tau}}(\frac{\sigma_j}{\rho_j})^{\frac{1}{4\tau s\mu_j^2}},
\end{cases}
\tag{7.8}
$$

where C is the smallest of the constants in Proposition 5 and Lemma 7.

The question is now to choose α_j and ρ_j so that, besides (7.9)=(6.8) and (7.10)=(6.9), also

$$
\begin{cases}
\varepsilon_j \leq \min[(\frac{1}{2s\beta_{j+1}\mu_{j+1}}\frac{\sigma_{j+1}}{\gamma_{j+1}})^{4s\mu_{j+1}^3}, \mid \mathcal{P}_{j+1}(\lambda_{j+1})\mid^{4s\mu_{j+1}^3}] \\
\rho_j \leq \min[\varepsilon_{j-1}^{\frac{1}{\mu_j}}, (\frac{\kappa}{8\mu_{j+1}\beta_j}\frac{1}{\lambda_j})^{2\tau s\mu_j^2}\sigma_j].
\end{cases}
\tag{7.11}
$$

Now λ_j increases at least as fast as 2^j, so if we define

$$
\alpha_j = \frac{1}{\lambda_j}
\tag{7.12}
$$

then α_j will decay exponentially fast. Since $\rho_j \leq \varepsilon_{j-1}^{\frac{1}{\mu_j}}$, ρ_j will also decay fast. Then the conditions are fulfilled if ρ_j satisfy

$$
\begin{cases}
\varepsilon_j \leq \min[(\frac{C\sigma_1}{\beta_1})^{(\mu_1\cdots\mu_{j+1})^6}, \rho_j^{(\mu_1\cdots\mu_{j+1})^6}, \mid \mathcal{P}_{j+1}(\lambda_{j+1})\mid^{4s\mu_{j+1}^3}] \\
\rho_j \leq \varepsilon_{j-1}^{\frac{1}{\mu_j}}.
\end{cases}
$$

If we let

$$
\rho_j^{(\mu_1\cdots\mu_{j+1})^6} = \varepsilon_j,
\tag{7.13}
$$

which defines ρ_j inductively, then these conditions amounts to

$$
\rho_j \leq \min[(\frac{C\sigma_1}{\beta_1}), \rho_{j-1}^{\frac{1}{(\mu_1\cdots\mu_j)^6\mu_j}}, \mid \mathcal{P}_{j+1}(\lambda_{j+1})\mid^{\frac{4s\mu_{j+1}^3}{(\mu_1\cdots\mu_{j+1})^6}}].
\tag{7.14}
$$

The first two conditions are easily seen to be fulfilled because ρ_j decays superexponentially. We now must discuss the partition and $\mathcal{P}_{j+1}(\lambda_{j+1})$.

Lemma 11. *We can choose the partition $\mathcal{P}'$ in Proposition 5 in such a way that the endpoints of the pieces of $\mathcal{P}'(N)$, for any N are contained in*

$$\cup\{T_a(\text{endpoints of } \mathcal{P}) : | a | \le (11\lambda + 2\nu' + \frac{1}{\kappa}(\frac{1}{\delta})^{\tau+1})n + N\},$$

where

$$\delta = (C\frac{\rho}{\beta\mu^2})^{2\mu}\frac{1}{\gamma}$$

and

$$n = \frac{1}{2\log 2}(\log((\nu' - 8\lambda)(\alpha - \alpha')) - \log\log(\frac{1}{\varepsilon})).$$

(When $\mathcal{P}$ has only one piece we can let an arbitrary point be its "endpoint".)

Proof. Due to the Diophantine condition (7.2) any interval of length δ will contain at least one element of the orbit $\{T_a x : | a | \le \frac{1}{\kappa}(\frac{1}{\delta})^{\tau+1}\}$ for any x, in particular for $x \in \mathcal{P}$. Hence we can choose such points as endpoints of the partition into pieces of length δ. Since

$$\mathcal{P}' = \mathcal{R}^n\mathcal{P}, \qquad \mathcal{R}\mathcal{P} = \mathcal{P}(5\lambda)_\delta(6\lambda + 2\nu'),$$

the result follows. $\square$

It follows that the partitions from Proposition 8 can be chosen so that the endpoints of $\mathcal{P}_j(\lambda_j)$ are contained in

$$\cup\{T_a(\text{endpoints of } \mathcal{P}_1) : | a | \le J_j\}$$

where

$$J_j = \lambda_j + \sum_{i=2}^{j}(11\lambda_{i-1} + 2\nu_i + \frac{1}{\kappa}(\frac{1}{\delta_{i-1}})^{\tau+1})n_i \le (\frac{1}{B\rho_{j-1}})^{4(\tau+1)\mu_{j-1}^2},$$

$$n_i = \frac{1}{2\log 2}(\log((\nu_i - 8\lambda_{i-1})(\alpha_{i-1} - \alpha_i)) - \log\log(\frac{1}{\varepsilon_{i-1}})),$$

where $B = B(\kappa, \tau, s, \dim \mathcal{L}, \beta_1, \gamma_1, \mu_1)$. This implies that any interval of length

$$\le \frac{\kappa}{J_j^\tau}$$

intersects at most $p = \#\mathcal{P}_1$ many pieces of the partition $\mathcal{P}_j(\lambda_j)$, i.e.

$$| \mathcal{P}_j(\lambda_j) | \ge \frac{\kappa}{J_j^\tau} \ge (\kappa B\rho_{j-1})^{4(\tau+1)\mu_{j-1}^2}.$$

The third condition on ρ_j in (7.14) therefore becomes

$$\rho_j \le (\kappa B\rho_j)^{\frac{4\tau(\tau+1)\mu_j^2 4s\mu_{j+1}^3+1}{(\mu_1\cdots\mu_{j+1})^6}}$$

which is fulfilled if for example $\mu_1^2 \ge 16s\tau(\tau + 1)$.

We can now derive the following theorem.

Theorem 12. *Let* $D \in \mathcal{NF}(\alpha,\ldots,\rho;\Omega,\mathcal{P})\&\mathcal{T}(\sigma,s)$ *be covariant with respect to the quasi-periodic $\mathcal{L}$-action (7.1-2), and assume that D is truncated at distance ν from the diagonal. Let F be a covariant matrix, smooth on $\mathcal{P}$ and*

$$| F_a^b |_{C^k} \le \varepsilon e^{-\alpha|b-a|}\gamma^k \qquad \forall k \ge 0.$$

Then there exists a constant C – C depends only on $\dim \mathcal{L}, \kappa, \tau, s, \alpha, \beta, \gamma, \lambda, \mu, \nu, \rho, \sigma, \#\mathcal{P}$ – such if $\varepsilon \le C$ then there exists a matrix U such that

$$U(x)^{-1}(D(x) + F(x))U(x) = D_\infty(x), \quad \forall x \in X,$$

and $D_\infty(x)$ is a norm limit of normal form matrices $D_j(x)$. Moreover $D_\infty(x)$ is pure point for a.e. x.

The limit $\lim_{j\to\infty} E_j(x) = E_\infty(x)$ is uniform and satisfies, for all $y \notin 2\pi\mathbb{Z}$,

$$Lebesgue\{x : E_\infty(x+y) - E_\infty(x) = 0\} = 0.$$

Moreover, if the E_j's and E_∞ are real, then for all subsets Y

$$Lebesgue(E_\infty^{-1}(Y)) = 0 \quad if \quad Lebesgue(Y) = 0.$$

If D and F are Hermitian, then U is unitary and D_∞ is Hermitian.

Proof. Define the sequences $\alpha_j, \beta_j, \gamma_j, \lambda_j, \mu_j, \nu_j, \rho_j, \sigma_j, \varepsilon_j$ by the formulas (7.8) and (7.12-13), starting with $\alpha, \beta, \gamma, \lambda, \mu, \nu, \rho, \sigma, \varepsilon$. Then they verify the assumptions (6.7-9), so we can apply Proposition 8, and (7.11), so we can apply Lemma 9-11.

Use now Proposition 8(I-II). By Lemma 9–11 we get $\mathcal{C}(\lambda_j, \mu_j, \nu_j)$-clustering for all $j \ge 1$, hence statement $(\mathcal{S}'_\infty)$ holds. The first part of the theorem now follows from Proposition 8(III) – it also gives the Hermitian case.

In order to prove that $D_\infty(x)$ is pure point for a.e. x we must estimate the number η_j in Proposition 8(IV). This is the only place where we use condition (6.4).

For each a in $|a| \le \nu_{j+1} + 2\lambda_j$ we consider

$$u_j(x, x + a\omega) = Res((D_j)_{\Omega_j(x+y)}(x + a\omega), (D_j)_{\Omega_j(x)}(x))$$

and the set

$$| u_j(x, x + a\omega) | \le \frac{\rho_j}{\beta_j^{\mu_j^2}}.$$

Using Lemma B1 we get that this this is a union of intervals of length less than

$$L_j = \frac{2}{\gamma_j}(2\rho_j \frac{(\nu_{j+1} + 2\lambda_j)^\tau}{\sigma_j\kappa})^{\frac{1}{\tau\mu_j^2}}.$$

The number of intervals does not exceed

$$M_j = 2^{s\mu_j^2}[\frac{\gamma_j(\nu_{j+1} + 2\lambda_j)\tau}{\kappa\sigma_j}8(\beta_j\mu_j^2)^{\mu_j^2}(s\mu_j^2 + 1)^2 + \#\mathcal{P})].$$

This gives an upper bound for η_j,

$$\eta_j \leq \text{const}(\nu_{j+1} + 2\lambda_j)^{\dim \mathcal{L}} L_j M_j$$

and the verification that

$$\sum \eta_j < \infty$$

is straight forward (with the choice we have made of ν_{j+1}).

The limit is uniform since

$$| E_{j+i}(x) - E_j(x) | \leq | D_{j+i}(x) - D_j(x) |.$$

Clearly,

$$\{x : E_\infty(x+y) - E_\infty(x) = 0\} \subset \{x : | E_j(x+y) - E_j(x) | \leq | D_\infty(x) - D_j(x) |\}$$

and this set is easy to estimate using the transversality on blocks of the matrix D_j.

Suppose now that the E_j's and E_∞ are real and let Y be a set of measure 0. We can use a covering $\cup_i I_i \supset Y$ with the following property: each I_i is contained in an interval $\tilde{I}_i$, and each $\tilde{I}_i$ is contained in a piece of the partition $\mathcal{P}_{j_i}(\lambda_{j_i})$, $j_i \geq J$, in such a way that its distance to the boundary of $\tilde{I}_i$ is at least

$$| D_\infty(x) - D_{j_i}(x) |.$$

Then E_{j_i} is smooth on $\tilde{I}_i$ and

$$\sum_i Lebesgue(E_\infty^{-1}(I_i)) \leq \sum_i Lebesgue(E_{j_i}^{-1}(\tilde{I}_i)).$$

This sum is easy to estimate – it is essentially equal to its first terms because of the rapid convergence – using the estimate of the resultants. Since we can do this for any J, this shows that the measure of $E_\infty^{-1}(Y)$ is 0 if the measure of Y is 0. $\square$

In the multi-level case the proofs goes through in the same way and all the results remain the same with obvious modifications.

We can now complete the diagram of this perturbative process in Fig. 3.

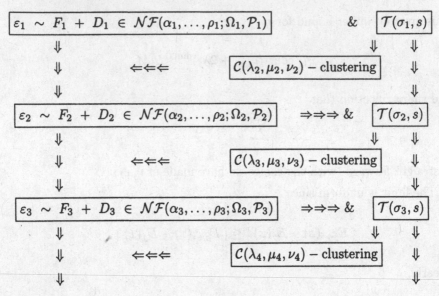

Figure 3

8 Applications

8.1 Discrete Schrödinger Equation

We consider the equation in strong coupling

$$-\varepsilon(u_{n+1} + u_{n-1}) + V(\theta + n\omega)u_n = Eu_n, \qquad n \in \mathbb{Z}, \qquad (8.1)$$

where V is a real valued function on the one-dimensional torus $\mathbb{T} = \mathbb{R}/(2\pi\mathbb{Z})$ and ω is a real number. We assume that V is piecewise Gevrey smooth on a partition with p many pieces and

$$|V|_{\mathcal{C}^k} \leq \beta\gamma^k \qquad \forall k \geq 0. \qquad (8.2)$$

The functions $V(\theta + y) - V(\theta))$ satisfies for all θ, y the two transversality conditions

$$\begin{cases} \max_{0 \leq k \leq s} |\partial_y^k(V(\theta + y) - V(\theta))| \geq \sigma \\ \max_{0 \leq k \leq s} |\partial_\theta^k(V(\theta + y) - V(\theta))| \geq \sigma \|y\|. \end{cases} \qquad (8.3)$$

We assume also that ω is Diophantine, i.e. for some $\tau > 1, \kappa > 0$

$$\|n\omega\| \geq \frac{\kappa}{|n|^\tau} \qquad \forall n \in \mathbb{Z} \setminus 0. \qquad (8.4)$$

Theorem 13. *Under the assumptions (8.2-4) there exist a small constant $\varepsilon_0(p, \beta, \gamma, s, \sigma, \kappa, \tau)$ and, for all $|\varepsilon| \leq \varepsilon_0$, a function $E_\infty(\theta)$ such that for almost all $\theta \in \mathbb{T}$ and for all $k \in \mathbb{Z}$, equation (8.1) has a solution $u^k \in l^2(\mathbb{Z})$ with $E = E_\infty(\theta + k\omega)$. The set $\{u^k : k \in \mathbb{Z}\}$ is an orthogonal basis for $l^2(\mathbb{Z})$.*

If we identify the right hand side of (8.1) as an operator H_θ, then the statement says that H_θ is pure point for almost all θ. The spectrum of H_θ is the closure of $Range(E_\infty)$ and it is easy to verify that the Lebesgue measure of $[\inf V, \sup V] \setminus Range(E_\infty)$ goes to 0 when ε goes to 0.

Proof. We identify (8.1) with a matrix $D + \varepsilon F$ as in the introduction. The matrix D is diagonal and

$$D \in \mathcal{NF}(\alpha = 1, \beta, \gamma, \lambda = 1, \mu = 1, \nu = 1, \rho = 1).$$

Since

$$\max_{0 \le k \le s\mu^2} \left| \frac{1}{(k!)^2 \gamma^k} \partial_y^k (V(\theta + y) - V(\theta)) \right| \ge \frac{\sigma}{(s!)^2 \gamma^s} = \tilde{\sigma},$$

and similar for the θ-derivative, we have also $D \in \mathcal{T}(\tilde{\sigma}, s)$. The perturbation εF satisfies

$$\left| (\varepsilon F)_a^b \right|_{C^k} \le (\varepsilon e) e^{-\alpha|a-b|} \qquad \forall k \ge 0.$$

The theorem is now an immediate consequence of Theorem 12. □

In weak coupling we consider the equation

$$-(u_{n+1} + u_{n-1}) + \varepsilon V(\theta + n\omega) u_n = E u_n, \qquad n \in \mathbb{Z}, \qquad (8.5)$$

where V is a real valued function on the torus $\mathbb{T}^d$ and $\omega \in \mathbb{R}^d$. We assume that V is analytic, satisfying

$$\sup_{|\Im\theta| < r} |V(\theta)| \le 1, \qquad (8.6)$$

and that ω is Diophantine, i.e. for some $\tau > d, \kappa > 0$

$$\|< k, \omega >\| \ge \frac{\kappa}{|k|^\tau} \qquad \forall k \in \mathbb{Z}^d \setminus 0. \qquad (8.7)$$

Theorem 14. *Under the assumptions (8.6-7) there exist a constant $\varepsilon_0(r, \kappa, \tau)$ and for all $|\varepsilon| \le \varepsilon_0$ a function $E_\infty(\xi)$ such that for almost all $\xi \in \mathbb{T}$ and for all $k \in \mathbb{Z}^d$, equation (8.5) has a solution*

$$u_n^k = e^{in(\xi + <k, \omega>)} U^k(\theta + n\omega)$$

in $l^\infty(\mathbb{Z})$ with $E = E_\infty(\xi + < k, \omega >)$. The set $\{U^k : k \in \mathbb{Z}^d\}$ is an orthogonal basis for $L^2(\mathbb{T}^d)$.

Proof. The equation for $U \in L^2(\mathbb{T}^d)$ is identified, in Fourier coefficients, with a matrix $D + \varepsilon F$ as in the introduction. The matrix D is diagonal and

$$D \in \mathcal{NF}(\alpha = r, \beta = 2, \gamma = \frac{1}{r}, \lambda = 1, \mu = 1, \nu = 1, \rho = 1).$$

The eigenvalue is $E(\xi) = \cos(\xi)$ and the functions $E(\xi + y) - E(\xi)$ satisfies the transversality conditions (8.3) for all ξ, y with some σ and $s = 2$. (Any non-constant analytic function with primitive period 2π satisfies condition (8.3) for some σ, s.) Hence $D \in \mathcal{T}(\tilde{\sigma}, 2)$. The Fourier coefficients of V decays exponentially with the factor $\alpha = r$ which shows that εF satisfies the required smallness condition. The theorem is now an immediate consequence of Theorem 12. □

These two theorems claims nothing about exponential decay of the solutions u^k or analyticity of the U^k, i.e. exponential decay of its Fourier coefficients. The uniform exponential decay of all the u^k's is measured in the iteration process by the sequence α_j which goes to 0, and must do so. If we instead let α vary over the partition so that we had one α-value for each block, it follows from the proof that $\alpha_j(x)$ stops decaying towards 0 as soon as the blocks $\Omega_j(x)$ stops increasing. This will give exponential decay. (Exponential decay of eigenvectors have been proven in [3,4,6].)

8.2 Discrete Linear Skew-Products

We consider first the *weakly perturbed* skew-product of the form

$$X_{n+1} - (A + \varepsilon B(\theta + n\omega))X_n = EX_n, \qquad n \in \mathbb{Z}, \tag{8.8}$$

where $A \in Gl(N, \mathbb{R})$, $B : \mathbb{T}^d \to gl(N, \mathbb{R})$. We assume that B is analytic, satisfying

$$\sup_{|\Im\theta| < r} |B(\theta)| \leq 1 \tag{8.9}$$

and that ω is Diophantine.

Theorem 15. *Under the assumptions (8.7) and (8.9) there exist a constant $\varepsilon_0(r, \kappa, \tau)$ and, for all $|\varepsilon| \leq \varepsilon_0$, functions $E_\infty^j(\xi)$, $j = 1, \ldots, N$, such that for almost all $\xi \in \mathbb{T}$ and for all $k \in \mathbb{Z}^d$, $j = 1, \ldots, N$, equation (8.8) has a solution*

$$X_n^{k,j} = e^{in(\xi + <k,\omega>)}Y^{k,j}(\theta + n\omega)$$

in $l^\infty(\mathbb{Z}) \otimes \mathbb{C}^N$ with $E = E_\infty^j(\xi + <k, \omega>)$. The set of functions $\{Y^{k,j} : k \in \mathbb{Z}^d, j = 1, \ldots, N\}$ is an orthogonal family in $L^2(\mathbb{T}^d) \times \mathbb{C}^N$.

Proof. Assume first that A is semi-simple. Since we construct complex-valued solutions we can without restriction assume that A and B are complex valued and that A is diagonal with diagonal elements $a_1, \ldots, a_N$.

The equation for $Y \in L^2(\mathbb{T}^d) \otimes \mathbb{C}^N$ is identified, in Fourier coefficients, with a matrix $D + \varepsilon F$ as in the introduction. The matrix D is diagonal and

$$D \in \mathcal{NF}(\alpha = r, \beta, \gamma = \frac{1}{r}, \lambda = 1, \mu = 1, \nu = 1, \rho = 1),$$

with $\beta =| A |$. It is now a multi-level matrix with eigenvalues

$$E^j(\xi) = e^{i\xi} - a_j, \qquad j = 1, \ldots, N$$

and the blocks $\Omega^j = \{0\} \times \{j\}, j = 1, \ldots, N$. The functions $E^i(\xi + y) - E^j(\xi)$ satisfy the transversality conditions (8.3) for all ξ, y with some σ and $s = 1$. Hence $D \in \mathcal{T}(\tilde{\sigma}, 1)$.

The Fourier coefficients of B decays exponentially with the factor $\alpha = r$ which shows that εF satisfies the required smallness condition. The theorem is now an immediate consequence of Theorem 12 – multi-level version.

The case when A is not semi-simple is almost the same. The matrix D is still on normal form,

$$D \in \mathcal{NF}(\alpha = r, \beta, \gamma = \frac{1}{r}, \lambda = N, \mu = N, \nu = 1, \rho = 1),$$

with $\beta =| A |$. It is still a multi-level matrix with eigenvalues

$$E^j(\xi) = e^{i\xi} - a_j, \qquad j = 1, \ldots, N$$

but with the blocks $\Omega^j = \{0\} \times \{1, \ldots, N\}, j = 1, \ldots, N$, reflecting that it is block diagonal instead of diagonal.

The transversality of the functions $E^i(\xi + y) - E^j(\xi)$ implies $D \in \mathcal{T}(\tilde{\sigma}, N^2)$ through Lemma B2. Now we can apply theorem 12. $\square$

The right conjecture about equation (8.8), under conditions (8.7,9), is the following: for $E = 0$ (or for any fixed E), if ε is small enough then (8.8) has a Floquet representation for almost all matrices A. This means that the fundamental solutions can be written as

$$e^{n\Delta} Z(\theta + n\omega), \qquad n \in \mathbb{Z},$$

where $Z : \mathbb{T}^d \to Gl(N, \mathbb{R})$ and $\Delta \in gl(N, \mathbb{R})$. Theorem 15 and its proof is probably not so far from proving this conjecture.

An even stronger result may be true: instead of "almost all matrices A" we may put "almost all matrices in any "generic" one-parameter family A_t". Such a result has been proven for certain one-parameter families in $Sl(2, \mathbb{R})$ [13] and in compact groups [17,18].

The application to the strongly perturbed skew-product (1.4) is straight forward, but the continuous Schrödinger in one dimension in strong coupling (1.5) is more delicate. Theorem 12 will not apply to the discrete equation (1.6), derived from (1.5) in the introduction, since W will not satisfy the full transversality condition. It seems however that W is transversal in a neighborhood of $W^{-1}(0)$ when E is close to $\inf V$, if V is transversal near $\inf V$. We therefore believe that when ε small enough and when E is sufficiently close to $\inf V$, then the spectrum of the operator (1.6) is pure point near 0 – this has been proven in some cases [4].

A   Appendix A

Gevrey classes

Let $u_j, j = 1, 2$ be smooth functions in the Gevrey class $\mathcal{G}_2$, defined in a convex set V in $\mathbb{R}^d$, and consider the norm

$$| u_j |_C =: \sup_{0 \le l \le k} \sup_{x \in V} | \frac{1}{(l!)^2} \partial^l u_j(x) |$$

– here k and l are multi-indices in $\mathbb{N}^d$.

Lemma A.1 *Assume that* $| u_j |_{C^k} \le \beta_j \gamma^k$, $\forall k \ge 0$. *Then for all* $k \ge 0$
   (i)

$$| u_1 u_2 |_{C^k} \le 4^d (\beta_1 \beta_2) \gamma^k;$$

   (ii)

$$| e^{u_1} |_{C^k} \le e^{4^d \beta_1} \gamma^k;$$

   (iii)

$$| \sqrt{1 + u_1} |_{C^k} \le (1 + \beta_1) \gamma^k \quad if \quad 4^d \beta_1 < 1;$$

   (iv)

$$| \frac{1}{u_1} |_{C^k} \le \frac{1}{\delta} (\frac{4^d \beta_1}{\delta} \gamma)^k \quad if \quad | u_1 | > \delta.$$

Proof.

$$| \partial^k (u_1 u_2) | = | \sum_{0 \le l \le k} \binom{k}{l} \partial^{k-l} u_1 \partial^l u_2 |$$

$$\le (\beta_1 \beta_2) \gamma^k (k!)^2 \sum_{0 \le l \le k} \binom{k}{l} \frac{(l!)^2 ((k-l)!)^2}{(k!)^2},$$

and the result follows since

$$\sum_{0 \le l \le k} \binom{k}{l}^{-1} = \prod_{i=1}^{d} [\sum_{0 \le l_i \le k_i} \binom{k_i}{l_i}^{-1}] \le 4^d$$

for all k – just notice that $\binom{k_i}{l_i} \ge 2^{l_i}$ if $l_i \le \frac{k_i}{2}$.

We use the power series expansions of e^u and $\sqrt{1 + u}$ which converges absolutely in $| u | < \infty$ and in $| u | < 1$ respectively, to prove the second and third estimate.

Notice that $\mid u_1 \mid > \delta$ implies that $\beta_1 > \delta$. By a scaling we can suppose that $\delta = 1$ and $\beta_1 \geq 1$. The fourth statement is obvious for $k = 0$ so we proceed by induction on $\mid k \mid$. Hence

$$\mid \partial^k(\frac{1}{u_1}) \mid \; = \mid -\frac{1}{u_1} \sum_{0 \leq l \leq k, \; l \neq 0} \binom{k}{l} \partial^{k-l}(\frac{1}{u_1}) \partial^l u_1 \mid$$

$$\leq \; 4^{d(\mid k \mid - 1)}(k!)^2 (\beta_1 \gamma)^k \sum_{0 \leq l \leq k} \binom{k}{l} \frac{(l!)^2((k-l)!)^2}{(k!)^2}$$

$$\leq \; (k!)^2 (4^d \beta_1 \gamma)^k.$$

□

Remark. The preceding lemma is valid also for matrices if we understand by $\frac{1}{u}$ the inverse of u, and if we use the operator norm satisfying

$$\mid uv \mid \leq \mid u \mid \mid v \mid.$$

Estimates of eigenvalues

Let $D(x)$ be a $m \times m$-matrix, smooth in a convex set V in $\mathbb{R}^d$ containing 0, and let $\{E^j(x)\}_1^m$ be a continuous choice of its eigenvalues.

Lemma A.2 *(i)*
$$\mid E^j(x) \mid \; \leq \; \mid D(x) \mid.$$

(ii)
$$\mid E^j(x) - E^j(x') \mid \; \leq \; 4 \mid D \mid_{C^0}^{1 - \frac{1}{m}} \mid \partial D \mid_{C^0}^{\frac{1}{m}} \mid x - x' \mid^{\frac{1}{m}}.$$

(iii) If D is Hermitian, i.e. $D^t = \bar{D}$, then

$$\mid E^j(x) - E^j(x') \mid \; \leq \; \mid \partial D(x) \mid_{C^0} \mid x - x' \mid.$$

Proof. (i) is obvious. To see (ii), let I be the segment joining x and x' and let $\delta = \mid E^1(x) - E^1(x') \mid$. Then there is an $x'' \in I$ such that,

$$\mid P(\lambda, x) \mid \; \geq \; (\frac{\delta}{4})^m, \quad \lambda = E^1(x''),$$

where $P(\lambda, x) = \prod_{i=1}^m (\lambda - E^j(x))$. (This is a well-known inequality of Chebyshev – see e.g. [28].) Hence

$$(\frac{\delta}{4})^m \; \leq \; \mid P(\lambda, x) - P(\lambda, x'') \mid \; \leq \; \mid \partial_x P(\lambda, \;) \mid_{C^0} \mid x - x'' \mid.$$

Since

$$\mid \partial_x P \mid \; \leq \; \sum_{j=1}^m \mid \det(D^1 \ldots \partial_x D^j \ldots D^m) \mid \; \leq \; \sum_{i,j=1}^m \mid \partial D_i^j \mid \mid D \mid^{m-1}$$

and $\sum | \partial D_i^j | \le m \, | \, \partial D \, |$ the result follows.

To see (iii) let $q^j(x)$ be the eigenvector corresponding to $E^j(x)$. Then

$$(D(x) - E^j(x)I)q^j(x) = 0,$$

and if we differentiate the relation and take the scalar product with $q^j(x)$ and use that the eigenvectors are orthogonal, then we get an estimate of ∂E^j. When E_j and q^j are not differentiable one uses the same argument with a difference operator. $\square$

Better estimates requires separation of eigenvalues. Assume that

$$| D |_{C^k} \le \beta \gamma^k \qquad \forall k \ge 0.$$

Lemma A.3 *Assume that the eigenvalues belong two groups*

$$\begin{cases} E^1(x), \dots, E^n(x) \\ E^{n+1}(x), \dots, E^m(x) \end{cases}$$

such that any eigenvalue of one group is separated from any eigenvalue of the other group by at least r for all $x \in V$. Then the polynomial

$$\prod_{j=1}^{n}(\lambda - E_j(x)) = \sum_{j=0}^{n} e_{n-j}(x)\lambda^j$$

is smooth in V and satisfies

$$| e_j |_{C^k} \le \binom{n}{j}\beta^j((\text{const}\frac{\beta m}{r})^m \gamma)^k \qquad \forall k \ge 0.$$

If D is Hermitian, then we also have

$$| \partial e_j |_{C^0} \le \binom{n}{j}\beta^j j\gamma.$$

The constants are independent of m.

Proof. Notice that $r \ge 2\beta$, and that the sup-estimate of the coefficients and, in the Hermitian case, of the first derivative of the coefficients follows from Lemma A.2. By scaling we can assume that $\beta = 1$ if we replace r by $\frac{r}{\beta}$.

The polynomial $P(\lambda, x) = \det(\lambda I - D(x))$ satisfies

$$| P(\lambda, \ ) |_{C^k} \le ((\text{const})^m \gamma)^k \qquad \forall k \ge 0,$$

if just $| \lambda | < 2$.

Choose a curve $\Delta(x)$ in $|\lambda| \leq 1 + \frac{1}{2}$ – piecewise constant in x – keeping a distance $\geq \frac{r}{2}$ to all the roots $E^1(x), \ldots, E^m(x)$ of $P(\lambda, x)$ and surrounding the first n of these roots. $\Delta(x)$ may consist of several components so we can choose it to be of length at most $n\pi r$. Then we have for all $\lambda \in \Delta(x)$

$$\begin{cases} |P(\lambda, x)| \geq (\frac{r}{2})^m = \rho \\ |\frac{\partial_\lambda P(\lambda, x)}{P(\lambda, x)}| \leq \frac{2m}{r}. \end{cases}$$

Using this we verify, as in Lemma A.1, that

$$|\frac{\partial_\lambda P(\lambda, x)}{P(\lambda, x)}|_{C^k} \leq \frac{2m}{r}((\text{const})^m \frac{1}{\rho}\gamma)^k \qquad \forall k \geq 0$$

– where we have used the Cauchy formula to estimate $|\partial_\lambda P(\lambda,)|_{C^k}$ on $|\lambda| \leq 1 + \frac{1}{2}$.

Consider now the power symmetric functions in the first n roots: $p_j(x) = E^1(x)^j + \cdots + E^n(x)^j$. We have the integral representation

$$p_j(x) = \frac{1}{2\pi} \oint_{\Delta(x)} \lambda^j \frac{\partial_\lambda P(\lambda, x)}{P(\lambda, x)} d\lambda,$$

from which we get

$$|p_j|_{C^k} \leq mn2^j((\text{const})^m \frac{1}{\rho}\gamma)^k \qquad \forall k \geq 0.$$

The e_j's are the elementary symmetric functions in the first n roots and we have the relation

$$e_j = \frac{(-1)^j}{j!} \det \begin{pmatrix} p_1 & 1 & 0 & 0 & \cdots \\ p_2 & p_1 & 2 & 0 & \cdots \\ p_3 & p_2 & p_1 & 3 & \cdots \\ \vdots & \vdots & \vdots & \vdots & \vdots \\ p_j & p_{j-1} & p_{j-2} & p_{j-3} & \cdots \end{pmatrix},$$

for $j \geq 1$ [29, page 20] – of course $e_0 = 1$. So $j!e_j$ is a sum of at most 2^j many products

$$p_{\iota_1} \cdots p_{\iota_l}, \qquad \iota_1 + \cdots + \iota_l = j \leq m,$$

each with a coefficient that is at most $(j-1) \cdot (j-2) \cdots l$. From Lemma A.1 we now get the estimate we want. $\square$

Orthogonalization

Let $v^1, \ldots, v^m$ be vectors in $\mathbb{C}^n$ which are smoothly defined in a convex neighborhood V of 0 in $\mathbb{R}^d$. Assume that $v^1(0), \ldots, v^m(0)$ is ON and that

$$|v^j|_{C^k} \leq \beta\gamma^k \qquad \forall k \geq 0,$$

for $j = 1, \ldots, m$. (This implies in particular that $\beta \geq 1$.)

Lemma A.4 *Then there is a constant, independent of m, and an upper triangular $m \times n$-matrix R, $R(0) = I$, smooth in*

$$W = \{x : | \, x \, | < \frac{const}{m\beta\gamma}\} \cap V,$$

such that

$$(w^1 \ldots w^m) = (v^1 \ldots v^m)R$$

is ON and for all $k \geq 0$

$$\begin{cases} | \, R \, |_{C^k} \leq (const\beta\gamma)^k \\ | \, w^j \, |_{C^k} \leq (const\beta\gamma)^k. \end{cases}$$

Proof. Since $v^1, \ldots, v^m$ is ON at $x = 0$ and has derivative bounded by $\beta\gamma$ it follows that, for $k \geq 0$,

$$\begin{cases} | < v^i, v^j > |_{C^k} \leq \frac{const}{m} & i \neq j \\ | < v^i, v^i > -1 \, |_{C^k} \leq \frac{const}{m} \end{cases}$$

in $| \, x \, | < \frac{const}{m\beta\gamma}$. In particular they remain linearly independent and of norm ≤ 2 if the constant is small enough.

Let now

$$\hat{v}^m = v^m - \sum_{i=1}^{m-1} a_i v^i$$

and determine the a_i's so that $< \hat{v}_m, v^i > = 0$ for all $i = 1, \ldots, m-1$. This gives a system of linear equation in the a_i's which we can solve by Gauss elimination if the constant is small enough (independent of m). The solution satisfies

$$\begin{cases} | \, a_i \, |_{C^k} \leq \frac{const}{m}\gamma^k & i = 1, \ldots, m-1 \\ | \, \hat{v}^m \, |_{C^k} \leq const\beta\gamma^k \end{cases}$$

for all $k \geq 0$.

If the constant is small enough we get

$$| \, \hat{v}^m \, | \geq \frac{1}{2}$$

and we can estimate $w^m = \frac{\hat{v}^m}{|\hat{v}^m|}$ by Lemma A.1. Then we proceed by induction. □

Subspaces and Angles

Assume now that Λ_1 and Λ_2 are two subspaces in $\mathbb{C}^m$ with $\Lambda_1 \cap \Lambda_2 = \{0\}$, and recall that they come equipped with ON-frames

Sublemma. *The spectrum of the matrices* $\Lambda_1^* \Lambda_2 \Lambda_2^* \Lambda_1$ *and* $\Lambda_2^* \Lambda_1 \Lambda_1^* \Lambda_2$ *only differ by an eigenvalue* 0 *of multiplicity* $= | \dim \Lambda_{\bar{1}} - \dim \Lambda_2 |$.

The spectrum of $I - \Lambda_1^* \Lambda_2 \Lambda_2^* \Lambda_1$ *is contained in* $[0, 1]$.

Proof. Notice that if $\lambda \neq 0$ is an eigenvalue of AB, then it is also an eigenvalue of BA, with the same multiplicity. Hence, the spectrum of AA^* and A^*A is the same except possibly for an eigenvalue 0. Now the kernel of A^*A is precisely the kernel of A, so if A is an $k \times l$-matrix of rank r then the dimension of the kernel is $l - r$. Since the rank of A and A^* is the same, we get that 0 is an eigenvalue of AA^* and A^*A of multiplicity $k - r$ and $l - r$ respectively.

If now $\Lambda_1 = \begin{pmatrix} I \\ 0 \end{pmatrix}$ and $\Lambda_2 = \begin{pmatrix} A \\ B \end{pmatrix}$, then

$$\Lambda_1^* \Lambda_2 \Lambda_2^* \Lambda_1 = AA^*, \quad \Lambda_2^* \Lambda_1 \Lambda_1^* \Lambda_2 = A^*A,$$

so the first first statement is proved.

Since $\Lambda_2^* \Lambda_2 = A^*A + B^*B$, we know that $I - A^*A = B^*B$. Now any matrix of the form B^*B is positive definite which implies that spectrum of $I - A^*A$ belongs to $[0, \infty[$ and to $]-\infty, 1]$. This proves the second statement. $\square$

We can now define the *angle* between Λ_1 and Λ_2 as the unique $\varphi \in [0, \pi[$ such that

$$\sqrt{dist(\sigma(\Lambda_1^* \Lambda_2 \Lambda_2^* \Lambda_1), 1)} = \sin(\varphi).$$

If these spaces are one-dimensional, this notion coincides with the "usual" angle between vectors.

Consider now an $m \times m$-matrix D, $| D | \leq \beta$, whose eigenvalues belongs to two groups that are separated by a distance at least r. Let Λ_1 and Λ_2 be the invariant subspaces corresponding to the two groups.

Lemma A.5 *The angle between the spaces* Λ_1 *and* Λ_2

$$\geq (const \frac{1}{m})^{\frac{m+1}{2}} (\frac{r}{\beta})^{(m+1)\frac{m}{2}},$$

if $r < const \beta$.

Proof. We can assume that

$$D = \begin{pmatrix} S & A \\ 0 & T \end{pmatrix},$$

where S and T are upper triangular with the diagonal elements of one separated by r from those of the other. We now look for an $m_1 \times m_2$-matrix R, $m = m_1 + m_2$, $m_2 \leq m_1$, such that

$$\begin{pmatrix} I & R \\ 0 & I \end{pmatrix}^{-1} D \begin{pmatrix} I & R \\ 0 & I \end{pmatrix} = \begin{pmatrix} S & 0 \\ 0 & T \end{pmatrix}.$$

Then the vectors

$$\begin{pmatrix} I \\ 0 \end{pmatrix} \quad \text{and} \quad \begin{pmatrix} R \\ I \end{pmatrix}$$

will span Λ_1 and Λ_2 respectively.

For a matrix M denote by M_i the matrix whose (k, l)-entry is M_k^l if $l - k = i$ and 0 otherwise. Then

$$S = S_0 + \cdots + S_{m_1}, \quad T = T_0 + \cdots + T_{m_2}, \quad A = A_{-m_1} + \cdots + A_{m_2}.$$

The equation for R,

$$SR - RT = -A,$$

can now be written

$$S_0 R_j - R_j T_0 = -A_j - \sum_{k \geq 1} S_k R_{j-k} + \sum_{l \geq 1} R_{j-l} T_l$$

for $-m_1 \leq j \leq m_2$. From this we get that

$$\mid R \mid \leq \text{const} m (\frac{\beta}{r})^{m+1}.$$

It follows from this by Gram-Schmidt that we get an ON-basis for Λ_2

$$\begin{pmatrix} A \\ B \end{pmatrix}$$

where B is a triangular $m_2 \times m_2$-matrix whose diagonal entries have absolute value

$$\geq \text{const} \frac{1}{m} (\frac{r}{\beta})^{m+1},$$

and $\mid B \mid$ satisfies the same estimate as $\mid R \mid$.

If now $v \in \mathbb{C}^{m_2}$ with $\mid v \mid = 1$, then

$$1 \geq \mid \begin{pmatrix} Av \\ Bv \end{pmatrix} \mid^2 = \mid Av \mid^2 + \mid Bv \mid^2,$$

which implies that

$$\mid A \mid^2 \leq 1 - [(\text{const} \frac{1}{m} (\frac{r}{\beta})^{m+1})^{m_2}]^2.$$

Since $\sqrt{\mid A^* A \mid} = \mid A \mid$, it follows that

$$\mid A^* A \mid \leq 1 - [(\text{const} \frac{1}{m} (\frac{r}{\beta})^{m+1})^{m_2}]^2,$$

which bounds the spectrum of $A^* A$. Since sine of the angle equals the square root of $dist(\sigma(A^* A), 1)$, the result follows. $\square$

Block splitting

Let D be an $m \times m$-matrix, smooth in a convex neighborhood V of 0 in $\mathbb{R}^d$, and assume that

$$| D |_{C^k} \leq \beta \gamma^k \qquad \forall k \geq 0.$$

Lemma A.6 *For any $r > 0$ there exists a $D(x)$-invariant decomposition*

$$\mathbb{C}^m = \sum_{i=1}^{k} \Lambda_j(x),$$

smooth on

$$W = \{| x | \leq (\text{const} \frac{r}{\beta m})^{2m} \frac{1}{\gamma}\} \cap V$$

with the following properties:
(i) all eigenvalues of D_{Λ_j} are r-close, i.e. for any given x

E *and* E' *are eigenvalues of the same block* $D(x)_{\Lambda_j(x)}$

$$\Rightarrow \qquad | E - E' | \leq r;$$

(ii)

$$| \Lambda_j |_{C^k} \leq ((\text{const} \frac{\beta m}{r})^{2m} \gamma)^k \qquad \forall k \geq 0;$$

(iii) the angles between the different Λ_j's are

$$\geq (\text{const} \frac{r}{\beta m})^{m^2};$$

(iv) the matrix $Q = (\Lambda_1 \ldots \Lambda_k)$ satisfies

$$| Q^{-1} |_{C^k} \leq (\text{const} \frac{\beta m}{r})^{2m^2} ((\text{const} \frac{\beta m}{r})^{4m^2} \gamma)^k \qquad \forall k \geq 0.$$

All the constants are independent of m.

Proof. We can take $\beta = 1$ if we replace r by $\frac{r}{\beta}$. Make now a decomposition of the eigenvalues of $D(0)$ into as many $\frac{r}{m}$-separated groups as possible, and consider the corresponding decomposition of $\mathbb{C}^m$ into invariant subspaces $\sum \Lambda_j(0)$. Choose a basis $v^1(0), \ldots, v^m(0)$ which is ON in each $\Lambda_j(0)$.

By Lemma A.2 the groups of eigenvalues of $D(x)$ remain $\frac{r}{2m}$-separated for x in

$$U = \{| x | \leq (\text{const} \frac{r}{\beta m})^m \frac{1}{\gamma}\} \cap V.$$

Let $\{E^j(x)\}_{j=1}^m$ be the eigenvalues of $D(x)$ and let $\check{P}_j(\lambda, x) = \prod(\lambda - E^i(x))$, where the product is over all eigenvalues except those associated to $\Lambda_j(0)$ – by Lemma A.3 we have estimates of the coefficients of $\check{P}_j(\lambda, x)$.

For $v^i(0)$ in $\Lambda_j(0)$, let

$$v^i(x) = \frac{1}{\check{P}_j(E^i(0),0)} \check{P}_j(D(x),x)v^i(0), \quad x \in U.$$

This gives a basis for $\Lambda_j(x)$ for x near 0 – how near? We have

$$| v^i |_{C^k} \le (\frac{m}{r})^{m-1} (\text{const})^m ((\text{const}\frac{m}{r})^m \gamma)^k \qquad \forall k \ge 0.$$

Hence, for $x \in W$, $v^i(x)$ will be close to $v^i(0)$ and we get a basis defined in W. Lemma A.4 gives an ON-basis for each $\Lambda_j(x)$ on W which fulfills the estimate.

The estimate of the angle follows from Lemma A.5, using that $m^2 \ge (m+1)\frac{m}{2}$.

In order to estimate the inverse of Q, consider the D^*-invariant decomposition

$$\mathbb{C}^m = \sum_{i=1}^{k} \tilde{\Lambda}_j(x),$$

such that the eigenvalues of $D^*(x)_{\tilde{\Lambda}_j(x)}$ are complex conjugates of the eigenvalues of $D(x)_{\Lambda_j(x)}$. Then $\tilde{\Lambda}_j$ satisfies the same estimate as Λ_j and $\tilde{\Lambda}_j = (\sum_{i \neq j} \Lambda_i)^\perp$.

If we let

$$\tilde{Q} = (\dots \tilde{\Lambda}_i \tilde{\Lambda}_j \dots)$$

then

$$B^{-1} =: \tilde{Q}^*Q = \begin{pmatrix} \ddots & & & \\ & (\tilde{\Lambda}_i)^*\Lambda_i & & \\ & & (\tilde{\Lambda}_j)^*\Lambda_j & \\ & & & \ddots \end{pmatrix}$$

so that $Q^{-1} = B\tilde{Q}^*$.

The angle between Λ_j and $(\sum_{i \neq j} \Lambda_i) = (\tilde{\Lambda}_j)^\perp$ is at least

$$(\text{const}\frac{r}{m})^{m^2}.$$

Therefore

$$S_j =: (\Lambda_j)^* \tilde{\Lambda}_j (\tilde{\Lambda}_j)^* \Lambda_j = I - (\Lambda_j)^*(\tilde{\Lambda}_j)^\perp((\tilde{\Lambda}_j)^\perp)^* \Lambda_j$$

is a symmetric matrix with spectrum bounded away from 0 by at least the square of that angle. Hence, by Lemma A.1,

$$| S_j^{-1} |_{C^k} \le (\text{const}\frac{m}{r})^{2m^2} ((\text{const}\frac{m}{r})^{4m^2} \gamma)^k \qquad \forall k \ge 0.$$

Since $((\tilde{\Lambda}_j)^*\Lambda_j)^{-1} = S_j^{-1}(\Lambda_j)^*\tilde{\Lambda}_j$ we get the same estimate for B as for S_j^{-1} and, hence, the same estimate for Q^{-1} as for S_j^{-1}. $\square$

Remark. When D is Hermitian in Lemma A.6 the angle between the invariant spaces is of course $\frac{\pi}{2}$ and Q is unitary so

$$| Q^{-1} |_{C^k} \le ((\text{const}\frac{\beta m}{r})^{2m}\gamma)^k \qquad \forall k \ge 0.$$

Truncation

Let

$$D : l^2(\mathcal{L}) \to l^2(\mathcal{L})$$

be an infinite-dimensional matrix with a basis of finite-dimensional (generalized) eigenvectors $\{q^b\}$ and blocks Ω^b. Suppose that D^* also has a basis of (generalized) eigenvectors $\tilde{q}^b$ with the same blocks Ω^b. We enumerate the eigenvectors so that q^b and $\tilde{q}^b$ correspond to complex conjugate eigenvalues with the same multiplicity. This implies that $\tilde{q}^b \perp q^a$, for all $a \ne b$.

Lemma A.7 *Let*

$$\Omega' = (\Omega^b)' = \cup_{\Omega^a \cap \Omega^b} \Omega^a, \qquad b \in \mathcal{L}.$$

If v is an (generalized) eigenvector of $D_{\Omega'}$ then either v is an (generalized) eigenvector of D or $v \perp \mathbb{C}^{\Omega^b}$.

Proof. Notice that q^a and $\tilde{q}^a$ are (generalized) eigenvectors of $D_{\Omega'}$ and $D_{\Omega'}^*$ respectively, if $\Omega^a \cap \Omega^b \ne \emptyset$.

Let now v be an (generalized) eigenvector of $D_{\Omega'}$. Then either $v = q^a$ for some a such that $\Omega^a \cap \Omega^b \ne \emptyset$ – and we are done – or $v \perp \tilde{q}^a$ for all a such that $\Omega^a \cap \Omega^b \ne \emptyset$. Since now $v = \sum A_c q^c$ we get

$$0 = <\tilde{q}^a, v> = A_a <\tilde{q}^a, q^a>,$$

which implies that $A_a = 0$. This means that

$$v = \sum_{\Omega^c \cap \Omega^b = \emptyset} A_c q^c$$

so v is perpendicular to $\mathbb{C}^{\Omega^b}$. $\quad\square$

A numerical lemma

Lemma A.8

$$\sum_{a_0=a,a_1,\ldots,a_j=b} e^{-(|a-a_1|+\cdots+|a_{j-1}-b|)s} \le (\sqrt{\dim\mathcal{L}}\frac{6}{s})^{(j-1)\dim\mathcal{L}} \qquad \forall a,b \in \mathcal{L}.$$

Proof. If we take $|\cdot|$ as the l^1-norm we get

$$\sum_{a_0=a,a_1,\ldots,a_j=b} e^{-(|a-a_1|+\cdots+|a_{j-1}-b|)s} =$$

$$\sum_{c_1,\ldots,c_{j-1}} e^{-(|(b-a)-c_1-\cdots-c_{j-1}|+|c_1|+\cdots+|c_{j-1}|)s} \leq$$

$$\left(\sum_c e^{-(|\frac{b-a}{j-1}-c|+|c|)\frac{s}{3}}\right)^{j-1} \leq \left(|\frac{b-a}{j-1}|+\frac{6}{s}\right)^{(j-1)d} e^{-|b-a|\frac{s}{3}} \leq \left(\frac{6}{s}\right)^{(j-1)d}$$

$(d = \dim \mathcal{L})$, where we used the inequality

$$|d - \sum_1^{j-1} c_i| + \sum_1^{j-1} |c_i| \geq \frac{1}{3}\sum_1^{j-1}(|\frac{d}{j-1} - c_i| + |c_i|).$$

This inequality is easy to verify because the $3\times$RHS is less then

$$\sum(|\frac{d}{j-1}|+2|c_i|) = |d|+2\sum|c_i| \leq$$

$$|d-\sum c_i| + |\sum c_i| + 2\sum|c_i| \leq |d - \sum c_i| + 3\sum|c_i|,$$

where summation goes from 1 to $j-1$.

The factor $\sqrt{\dim\mathcal{L}}$ comes in because we use th l^2-norm. $\square$

B Appendix B

Estimates of preimages

Let u be a real or complex valued smooth function defined on an open interval Δ in $\mathbb{R}$ and satisfying

$$\begin{cases} |u|_{C^k} \leq \beta\gamma^k & \forall\, 0 \leq k \leq m+1 \\ \max_{0\leq k\leq m} |\frac{1}{(k!)^2\gamma^k}\partial^k u(x)| \geq \sigma & \forall x \in \Delta. \end{cases}$$

Lemma B.1 *There exists a finite union of open intervals $\cup_{j\in J}\Delta_j$ such that*

$$\begin{cases} \#J & \leq 2^m[\frac{8\beta\gamma(m+1)^2}{\sigma}\,|\Delta|+1] \\ \max_{j\in J}|\Delta_j| & \leq \frac{2}{\gamma}(\frac{2\rho}{\sigma})^{\frac{1}{m}} \\ |u(x)| & \geq \rho \quad \forall x \in \Delta\setminus\cup\Delta_j. \end{cases}$$

Proof. Assume for simplicity that $0 \in \Delta$ and that $|\partial^m u(0)| \geq \sigma(m!)^2\gamma^m$. Then there is an interval $\tilde{\Delta}$ of length

$$\frac{\sigma}{8\beta\gamma(m+1)^2}$$

such that $|\partial^m u(x)| \geq \frac{\sigma}{2}(m!)^2\gamma^m$ on $\tilde{\Delta}$, if u is real – if u is complex valued then this holds for $\Re u$ or $\Im u$.

Consider now $\partial^{m-1}u$ on $\tilde{\Delta}$. There exists an interval Δ_1 of length

$$\geq \frac{2}{\gamma}(\frac{2\rho}{\sigma})^{\frac{1}{m}}$$

such that

$$|\partial^{m-1}u(x)| \geq \frac{\sigma}{2}(\frac{2\rho}{\sigma})^{\frac{1}{m}}((m-1)!)^2\gamma^{m-1} \qquad \forall x \in \tilde{\Delta} \setminus \Delta_1.$$

Consider now $\partial^{m-2}u$ on $\tilde{\Delta} \setminus \Delta_1$. There exist two intervals Δ_2, Δ_3, each of length

$$\frac{2}{\gamma}(\frac{2\rho}{\sigma})^{\frac{1}{m}},$$

such that

$$|\partial^{m-2}u(x)| \geq \frac{\sigma}{2}(\frac{2\rho}{\sigma})^{\frac{2}{m}}((m-2)!)^2\gamma^{m-2} \qquad \forall x \in \tilde{\Delta} \setminus \Delta_1 \cup \Delta_2 \cup \Delta_3,$$

etc.

Hence we obtain in $\tilde{\Delta}$, $2^m - 1$ (possibly void) intervals Δ_i such that

$$|u(x)| \geq \rho \qquad \forall x \in \tilde{\Delta} \setminus \cup \Delta_i,$$

$$|\Delta_i| \leq \frac{2}{\gamma}(\frac{2\rho}{\sigma})^{\frac{1}{m}}.$$

On the whole interval Δ we get at most

$$2^m \times [\frac{8\beta\gamma(m+1)^2}{\sigma}|\Delta|+1]$$

many such intervals. $\square$

Transversality of products of functions

Let now u_j be a sequence of real or complex valued smooth functions defined on an open interval Δ in $\mathbb{R}$ and satisfying

$$\begin{cases} |u_j|_{C^k} \leq \beta\gamma^k & \forall\, 0 \leq k \leq m+1 = m_1 + \cdots + m_j + 1 \\ \max\limits_{0 \leq k \leq m_i} |\frac{1}{(k!)^2\gamma^k}\partial^k u_j(x)| \geq \sigma & \forall x \in \Delta. \end{cases}$$

Lemma B.2 *If* $u = u_1 \cdots u_j$, *then*

$$\begin{cases} |u|_{C^k} \leq 4^{d(j-1)}\beta^j\gamma^k & \forall\, 0 \leq k \leq m+1, \\ \max\limits_{0 \leq k \leq m} |\frac{1}{(k!)^2\gamma^k}\partial^k u(x)| \geq (\frac{1}{2})^{8m^2}\sigma^{j(m+1)} & \forall x \in \Delta. \end{cases}$$

Proof. The first part follows from Lemma A.1 so we concentrate on the second estimate.

We can assume that $\beta = 1$ and $\gamma = 1$ if we replace the u_i's by $\frac{u_i}{\beta}$ and rescale x. Fix an $x - x = 0$ say. We can assume without restriction that

$$\frac{1}{(m_i!)^2} \mid \partial^{m_i} u_i(0) \mid \geq \sigma \qquad \forall i.$$

Order the m_i's in decreasing order $m_1 \geq m_2 \geq \ldots$.

Fix a symmetric interval I around 0 of length

$$2\delta \leq \frac{\sigma}{8(m+1)^2}.$$

By Lemma B1 we get that $\mid u_1 \mid \geq \rho$ outside a subset of I of measure less than

$$N(\frac{2\rho}{\sigma})^{\frac{1}{m_1}}, \qquad N = 2^{m_1+2},$$

which is less than $\frac{\delta}{j}$ if we chose

$$\rho = \frac{\sigma}{2}(\frac{\delta}{jN})^{m_1}.$$

Applying this to all the u_i's we get that

$$\mid u(x) \mid \geq \rho^j = (\frac{\sigma}{2})^j(\frac{\delta}{jN})^m$$

outside a set of measure less than δ.

Consider now the Taylor expansion

$$u(x) = a_0 + a_1 x + \cdots + a_m x^m + \hat{u}(x).$$

Then, for $\mid x \mid \leq \delta$,

$$\mid \hat{u}(x) \mid \leq (m+1)! \mid x \mid^{m+1} \leq (m+1)! \delta^{m+1}.$$

From this we get an $\frac{\delta}{2} \leq \mid x \mid \leq \delta$ such that

$$\mid a_0 + a_1 x + \cdots + a_m x^m \mid \geq \frac{1}{2}(\frac{\sigma}{2})^j(\frac{\delta}{jN})^m,$$

if we choose

$$\delta = \frac{1}{2(m+1)!}(\frac{\sigma}{2})^j(\frac{1}{jN})^m.$$

Hence for δ so small and some $\frac{\delta}{2} \leq \mid x \mid \leq \delta$ and some $0 \leq k \leq m$,

$$\mid a_k x^k \mid \geq \frac{1}{2(m+1)}(\frac{\sigma}{2})^j(\frac{\delta}{jN})^m.$$

Since $k! a_k = \partial^k u(0)$ the result follows. $\quad\square$

Remark. In [5] there was given a short proof of this lemma with a worse estimate, which however is good enough for the applications we have in mind. The above proof, which is just as short but gives a considerable better estimate, is due to M. Goldstein [6].

Acknowledgement. This paper was partially written during a visit to IMPA, Rio de Janeiro, financed by the STINT foundation.

References

1. Craig, W.: Pure point spectrum for discrete almost periodic Schrödinger operators. Commun. Math. Phys. **88**, 113–131 (1983)
2. Pöschel, J: Examples of discrete Schrödinger operators with pure point spectrum. Commun. Math. Phys. **88**, 447–463 (1983)
3. Sinai, Ya.G.: Anderson localization for the one-dimensional difference Schrödinger operator with a quasi-periodic potential. J. Stat. Phys. **46** 861–909 (1987)
4. Fröhlich, J., Spencer, T., Wittver, P.: Localization for a class of one-dimensional quasi-periodic Schrödinger operators. Commun. Math. Phys. **132**, 5–25 (1990)
5. Eliasson, L.H.: Discrete one-dimensional quasi-periodic Schrödinger operators with pure point spectrum. Acta Math. **179**, 153–196 (1997)
6. Goldstein, M.: Anderson localization for quasi-periodic Schrödinger equation. Preprint, 1998.
7. Chulaevsky, V.A., Dinaburg, E.I.: Methods of KAM-Theory for Long-Range Quasi-Periodic Operators on $\mathbb{Z}^\nu$. Pure Point Spectrum. Comm. Math. Phys. **153**, 559–577 (1993)
8. Avron,J., Simon, B.: Singular continuous spectrum for a class of almost periodic Jacobi matrices. Bull. Amer. Math. Soc. **6**, 81–85 (1982)
9. F. Delyon, D. Petritis: Absence of localization in a class of Schrödinger operators. Commun. Math. Phys. **103**, 441–443 (1986)
10. Dinaburg, E.I., Sinai, Ya.G.: The one-dimensional Schrödinger equation with quasi-periodic potential. Funkt. Anal. i. Priloz. **9**, 8–21 (1975)
11. Rüssmann, H.: On the one-dimensional Schrödinger equation with a quasiperiodic potential.In: *Nonlinear Dynamics* (Internat. Conf., New York, 1979), Helleman, R.H.G. (ed.) New York: New York Acad. Sci., 1980
12. Moser, J., Pöschel, J.: An extension of a result by Dinaburg and Sinai on quasi-periodic potentials. Comment. Math. Helvetici **59**, 39–85 (1984)
13. Eliasson, L.H.: Floquet solutions for the one-dimensional quasi-periodic Schrödinger equation. Commun. Math. Phys. **146**, 447–482 (1992)
14. Bellissard, J., Lima, R., Testard, D.: A metal insulator transition for the almost Mathieu equation. Commun. Math. Phys. **88**, 207–234 (1983)
15. Albanese, C.: KAM theory in momentum space and quasiperiodic Schrödinger operators. Ann. Inst. H. Poincaré, Anal. Non Linéaire **10**, 1–97 (1993)
16. Goldstein, M.: Laplace transform methods in the perturbation theory of spectrum of Schrödinger operators I and II. Preprints, 1991–92
17. Krikorian, R.: Réductibilité presque partout des systèmes quasi périodiques analytiques dans le cas $SO(3)$. C. R. Acad. Sci. Paris **321**, Série I, 1039–1044 (1995)

18. Krikorian, R.: *Réductibilité des systèmes produits croisés quasi-périodiques à valeurs dans des groupes compacts*. Paris: Thesis École Polytechnique, 1996
19. Eliasson, L.H.: Ergodic skew-systems on $SO(3, \mathbb{R})$. Preprint ETH-Zürich, 1991
20. Eliasson, L.H.: Reducibility, ergodicity and point spectrum. Proceedings of the International Congress of Mathematicians. Berlin 1998, Vol. II, 779–787 (1998)
21. Young, L.S.: Lyapunov exponents for some quasi-periodic cocycles. Ergod. Th.& Dynam. Syst. **17**, 483–504 (1997)
22. Chualevsky, V.A., Sinai, Ya.G.: *Anderson localization and KAM-theory*. In: P. Rabinowitz, E. Zehnder (eds.): Analysis etcetera. New York: Academic Press, 1989
23. Chualevsky, V.A., Sinai, Ya.G.: The exponential localization and structure of the spectrum of 1D quasi-periodic discrete Schrödinger operators. Rev. Math. Phys. **3**, 241–284 (1991)
24. Goldstein, M.: Quasi-periodic Schrödinger equation and analogue of the Cartan's estimate for real algebraic functions. Preprint, 1998
25. Eliasson, L.H.: *Communication at the 4th Quadriennial International Conference on Dynamical Systems*, IMPA, July 29–August 8, 1997
26. Bourgain, J.: Quasi-periodic solutions of Hamiltonian perturbations of 2D linear Schrödinger equation. Ann. of Math. **148**, 363–439 (1998)
27. Folland, G.D.: *Introduction to partial differential equations*. Princeton, N.J.: Princeton University Press, 1976
28. Shapiro, H.: *Topics in approximation theory*, LMN 187. Berlin: Springer-Verlag, 1971
29. MacDonald, I.G.: *Symmetric functions and Hall polynomials*. Oxford: Clarendon Press, 1979

KAM-persistence of finite-gap solutions

Sergei B. Kuksin

Introduction

The paper is devoted to a proof of the following KAM theorem: most of space-periodic finite-gap solutions of a Lax-integrable Hamiltonian PDE persist under small Hamiltonian perturbations as time-quasiperiodic solutions of the perturbed equation. In order to prove it we obtain a number of results, important by itself: a normal form for a nonlinear Hamiltonian system in the vicinity of a family of lower-dimensional invariant tori, etc. The paper is an abridged version of the manuscript [KK], missing proofs can be found there. Still our presentation is rather complete modulo a KAM-theorem for perturbations of linear equations (for its proof see [K, K2, P]).

Notations. Everywhere in the paper "linear map" means "bonded linear map". For a linear operator J we denote by $\bar{J}$ the operator $-J^{-1}$. For a complex Hilbert space, $\langle \cdot, \cdot \rangle$ stands for a complex-bilinear quadratic form such that $\langle u, \bar{u} \rangle = \|u\|^2$. We systematically use the following non-standart notations: for a vector $V = (V_1, \ldots, V_n) \in \mathbb{N}^n$ we denote $\mathbb{N}_V = \{m \in \mathbb{N} \mid m \neq V_j \; \forall j\}$ and $\mathbb{Z}_V = \{m \in \mathbb{Z} \mid m \neq \pm V_j \; \forall j\}$; we abbreviate $\mathbb{N}_{(1,2,\ldots,n)}$ to $\mathbb{N}_n$, etc. In particular, $\mathbb{Z}_0$ stands for the set of non-zero integers.

1 Some analysis in Hilbert spaces and scales

1.1 Smooth and analytic maps

Below we shall work with differentiable maps between domains in Hilbert (more general, Banach) spaces. Since the category of C^r-smooth Fréchet maps with $r \geq 2$ is rather cumbersome and since only analytic object arise in our main theorems, we mostly restrict ourselves to the two extreme cases: with few exceptions the maps will be either C^1-smooth or analytic. Now we fix corresponding notations and briefly recall some properties of C^1-smooth and analytic maps.

Let X, Y be Hilbert spaces and O be a domain in X. A continuous map $f : O \to Y$ is called continuously differentiable, or C^1-smooth (in the sense of Fréchet) if there exists a bounded linear map $f_*(x) : X \to Y$ which continuously depends on $x \in O$, such that $f(x + x_1) - f(x) = f_*(x)x_1 + o(\|x_1\|_X)$ provided that $x, x + x_1 \in O$. We call $f_*(x)$ a *derivative* of f or its *tangent map*. By $f^*(x)$ we denote the adjoint map $f^*(x) = (f_*(x))^* : Y \to X$.

For a real Hilbert space X we denote by X^c its *complexification*, $X^c = X \otimes_{\mathbb{R}} \mathbb{C}$.

Examples. If X is an L_2-space or a Sobolev space of real-valued functions, then X^c is a corresponding space of complex functions. If X is an abstract separable Hilbert space and $\{\phi_j\}$ is its Hilbert basis, then $X = \{\sum x_j\phi_j \mid x_j$'s are real and $\sum |x_j|^2 < \infty\}$, while $X^c = \{\sum z_j\phi_j \mid z_j$'s are complex and $\dots\}$.

Let X^c, Y^c be complex Hilbert spaces and O^c be a domain in X^c. A map $f : O^c \to Y^c$ is called Fréchet-analytic if it is C^1-smooth in the sense of real analysis (when we treat X^c, Y^c as real spaces) and the tangent maps $f_*(x)$ are complex-linear. Locally near any point in O^c such a map can be represented as a normally convergent series of homogeneous maps (see [PT]).

For real Hilbert spaces X, Y and a domain $O \subset X$, a map $F : O \to Y$ is analytic if it can be extended to a complex-analytic map $F : O^c \to Y^c$, where O^c is a complex neighbourhood of O in X^c.

A map $F : X \supset O \to Y$ is called δ-analytic (δ is a positive real number) if it extends to a bounded analytic map $(O + \delta) \to Y^c$ ($O + \delta$ is the δ-neighbourhood of O in X^c).

We note that compositions of analytic maps are analytic, as well as their linear combinations. Besides, any analytic map is C^k-smooth for every k.

There is an important *criterion of analyticity*: a map $f : X^c \supset O^c \to Y^c$ is analytic if and only if it is locally bounded [1] and *weakly analytic*, i.e., for any $y \in Y^c$ and any affine complex plane $\Lambda \subset X^c$ the complex function $\Lambda \cap O^c \to \mathbb{C}, \lambda \mapsto \langle F(\lambda), y \rangle_Y$ is analytic in the sense of one complex variable. Even more, it is sufficient to check analyticity of these functions for a countable system $y = y_1, y_2, \dots$ of vectors in Y such that the linear envelope of this system is dense in Y (see [PT]).

The *Cauchy estimate* states that if a map $F : X^c \supset O^c \to Y^c$ admits a bounded analytic extension to $O^c + \delta$, then for any $u \in O^c$ one has:

$$\|F_*(u)\|_{X,Y} \le \delta^{-1} \sup_{u' \in O^c + \delta} \|F(u')\|_Y.$$

(The estimate readily follows from its one-dimensional version applied to the holomorphic functions $O_\delta(\mathbb{C}) \ni \lambda \mapsto \langle F(u + \lambda x), y \rangle_Y$, where $\|x\|_X = \|y\|_Y = 1$). In particular, this estimate applies to δ-analytic maps between subsets of real Hilbert spaces.

If $F : X^c \supset O^c \to Y^c$ is an analytic map and for some point $x \in O^c$ the tangent map $F_*(x)$ is an isomorphism, then by the *inverse function theorem* in a sufficiently small neighbourhood of x the map F can be analytically inverted. The same is true for real analytic maps. See [PT].

For *Banach spaces* everything is much the same with one extra difficulty: there is no canonical way to define a complexification X^c of a real Banach space X. This difficulty should not worry us since all Banach spaces used

---

[1] that is, any point $x \in O^c$ has a neighbourhood, where f is uniformly bounded. In particular, any continuous map is locally bounded.

below are natural and one can immediately guess the right complexification. For example, if X is the space of bounded linear operators $Y_1 \to Y_2$ where Y_1, Y_2 are Hilbert spaces, then X^c is the complex space of linear over reals operators $Y_1 \to Y_2^c$, etc.

1.2   Scales of Hilbert spaces and interpolation

Below we shall study one-dimensional in space partial differential equations in appropriate *scales of Sobolev spaces* $\{X_s\}$, where $s \in \mathbb{Z}$ or $s \in \mathbb{R}$. That is,

(i) each X_s is a Hilbert subspace of a Sobolev space of order s, formed by scalar- or vector-functions on a segment $[0, T]$ (so $X_0 \subset L_2$);

(ii) there exist:

   (1) a Hilbert basis $\{\phi_k \mid k \in \mathbb{Z}_0\}$, $\mathbb{Z}_0 = \mathbb{Z} \setminus \{0\}$, of the space X_0,
   (2) an even positive sequence $\{\vartheta_j \mid j \in \mathbb{Z}_0\}$ of linear growth, i.e. $\vartheta_j = \vartheta_{-j}$ and $C^{-1}|k| \leq \vartheta_k \leq C|k|$ for all k,

such that for any real s the set of vectors $\{\phi_k \vartheta_k^{-s} \mid k \in \mathbb{Z}_0\}$ form a Hilbert basis of the space X_s.

The second assumption implies that for any $-\infty < a < b < \infty$ the space X_c, $c = (1 - \theta)a + \theta b$, *interpolates* the spaces X_a and X_b: in notations of [LM], $X_c = [X_a, X_b]_\theta$. In particular, for any $u \in X_b$ holds the *interpolation inequality*: $\|u\|_c \leq \|u\|_a^{1-\theta} \|u\|_b^\theta$. The basis $\{\phi_k\}$ is called *a basis of the scale*.

The norm and the scalar product in X_s will be denoted $\|\cdot\|_s$ and $\langle \cdot, \cdot \rangle_s = \langle \cdot, \cdot \rangle_{X_s}$; we abbreviate $\langle \cdot, \cdot \rangle_0 = \langle \cdot, \cdot \rangle$. Due to ii),

$$\langle u, u \rangle_s^2 = \|u\|_s^2 = \sum |u_k|^2 \vartheta_k^{2s} \quad \text{if} \quad u = \sum u_k \phi_k.$$

The scalar product $\langle \cdot, \cdot \rangle$ extends to a bilinear pairing $X_s \times X_{-s} \to \mathbb{R}$. For any space X_s (real or complex) we identify its adjoint $(X_s)^*$ with the space X_{-s}.

We denote by $X_{-\infty}$, X_∞ the linear spaces $X_{-\infty} = \cup X_s$, $X_\infty = \cap X_s$. The space X_∞ is dense in each X_s. It is formed by smooth (vector-) functions.

We shall often treat scales $\{X_s\}$ as abstract Hilbert scales and do not discuss their embeddings to a scale of Sobolev functions.

Example 1.1. Let $X_s = H_0^s(S^1, \mathbb{R})$ be the Sobolev space of 2π-periodic functions with zero mean-value. This scale satisfies (i)–(iii). For a basis $\{\phi_k \mid k \in \mathbb{Z}_0\}$ we take the trigonometric basis:

$$\varphi_k = \frac{1}{\sqrt{\pi}} \cos kx, \quad \varphi_{-k} = -\frac{1}{\sqrt{\pi}} \sin kx \quad \text{for} \quad k = 1, 2, \ldots \qquad (1.1)$$

(the minus-sign is introduced for further purposes). For a sequence ϑ_k we take $\vartheta_k = |k|$. This choise corresponds to the homogeneous scalar product in X_s, $\langle u, v \rangle_s = \int (u^{(s)} v^{(s)}) dx$. Complexification X_s^c of this space is the space $H_0^s(S^1; \mathbb{C})$ of complex Sobolev functions.

Given two scales $\{X_s\}$, $\{Y_s\}$ as above and a linear map $L : X_\infty \to Y_{-\infty}$, we denote by $\|L\|_{s_1,s_2} \leq \infty$ its norm as a map $X_{s_1} \to Y_{s_2}$. We say that the map L defines a *morphism of order d of the scales* $\{X_s\}$ *and* $\{Y_s\}$ for $s \in [s_0, s_1]$, if $\|L\|_{s,s-d} < \infty$ for each $s \in [s_0, s_1]$ with some fixed $-\infty \leq s_0 \leq s_1 \leq +\infty$. [2] If in addition the inverse map L^{-1} exists and defines a morphism of order $-d$ of the scales $\{Y_s\}, \{X_s\}$ for $s \in [s_0 + d, s_1 + d]$, we say that L defines an *isomorphism of order d* of the two scales. If $\{Y_s\} = \{X_s\}$, then an isomorphism L is called an *automorphism*. We shall drop the specification "for $s \in [s_0, s_1]$" if the segment $[s_0, s_1]$ is fixed for a moment or can be easily recovered.

If $L : X_s \to Y_{s-d}$ is a morphism of order d for $s \in [s_0, s_1]$, then the adjoint maps $L^* : (Y_{s-d})^* = Y_{-s+d} \to (X_s)^* = X_{-s}$ form a morphism of the scales $\{Y_s\}$ and $\{X_s\}$ of the same order d for $s \in [-s_1 + d, -s_0 + d]$. We call it the *adjoint morphism*.

A morphism L of a Hilbert scale $\{X_s\}$, complex or real, is called *symmetric (anti symmetric)* if $L = L^*$ (respectively $L = -L^*$) on the space X_∞. In particular, a linear operator $L : X_{s_0} \to Y_{s_0-d}$ is called symmetric (anti symmetric) if $L = L^*$ ($L = -L^*$) on the space X_∞.

If L is a symmetric morphism of $\{X_s\}$ of order d for $s \in [s_0, d - s_0]$, then L^* also is a morphism of order d for $s \in [s_0, d - s_0]$ and $L = L^*$ as the scale's morphisms. We call such a morphism *selfadjoint*. *Anti selfadjoint* morphisms are defined similar.

Example. The operator $-\triangle$ defines a selfadjoint automorphism of order 2 of the Sobolev scale $\{H_0^s\}$. The operator $\partial/\partial x$ defines an anti selfadjoint automorphism of order one.

Linear maps from one Hilbert scale to another obey the Interpolation Theorem:

Theorem 1 (see [LM, RS2].) *Let* $\{X_s\}$, $\{Y_s\}$ *be two real Hilbert scales and* $L : X_\infty \to Y_{-\infty}$ *be a linear map such that* $\|L\|_{a_1,b_1} = C_1$, $\|L\|_{a_2,b_2} = C_2$. *Then for any* $\theta \in [0,1]$ *we have* $\|L\|_{a,b} \leq C_\theta$, *where* $a = a_\theta = \theta a_1 + (1 - \theta)a_2$, $b = b_\theta = \theta b_1 + (1-\theta)b_2$ *and* $C_\theta = C_1^\theta C_2^{1-\theta}$. *This result with* C_θ *replace by* $4C_\theta$ *remain true for complex Hilbert scales.*

In particular, if under the theorem's assumptions $a_1 - b_1 = a_2 - b_2 =: d$, then L defines a morphism of order d of the scales $\{X_s\}$, $\{Y_s\}$ for $s \in [a_1, a_2]$.

Corollary. *Let L be a selfadjoint or an anti selfadjoint morphism of a scale* $\{X_s\}$ *such that* $\|L\|_{a,b} = C < \infty$ *for some a, b. Then* $\|L\|_{\theta(a+b)-b,\theta(a+b)-a} \leq C$ *for any* $0 \leq \theta \leq 1$.

Proof. Since $(X_a)^* = X_{-a}$ and $(X_b)^* = X_{-b}$, then $C = \|L\|_{a,b} = \|L^*\|_{-b,-a} = \|L\|_{-b,-a}$. Now the assertion follows from the Interpolation Theorem. □

---

[2] if $s_0 = -\infty$, then $s > s_0$ since $X_{-\infty}$ and $Y_{-\infty}$ are given no norms. Similar $s < \infty$ if $s_1 = \infty$.

In particular, an operator L as above defines a morphism of order $a - b$ of the scale $\{X_s\}$ for $s \in [-b, a]$ (or $\in [a, -b]$ if $-b > a$).

Let $-\infty < a \le b \le \infty$ and $O_s \subset X_s$, $s \in [a, b]$, be a system of compatible domains (i.e., $O_{s_1} \cap O_{s_2} = O_{s_2}$ if $s_1 \le s_2$). Let $F : O_a \to Y_{a-d}$ be an analytic (or C^1-smooth) map such that its restriction to the domains O_s with $a \le s \le b$ define analytic (or C^1-smooth) maps $F : O_s \to Y_{s-d}$. [3] Then we say that F is an *analytic (or C^1-smooth) map of order d* for $a \le s \le b$.

Example 1.1, continuation. Let us denote by Π the projector $\Pi : H^s \to H_0^s$, which sends a function $u(x)$ to $u(x) - \int u(x)\, dx/2\pi$. The Sobolev spaces H^s with $s > 1/2$ are Banach algebras: $\|uv\|_s \le C_s\|u\|_s\|v\|_s$, see [Ad]. Therefore for any segment $[a, b]$, $1/2 < a \le b \le \infty$, the map $u(x) \mapsto \Pi \circ F(u(x))$ where F is a polynomial, defines an analytic map $H_0^s \to H_0^s$ of order zero for $s \in [a, b]$. If $g(x)$ is any fixed function, then the map $u(x) \mapsto \Pi(F(u(x)) + g(x))$ is analytic of order zero for $s \in [a, b]$ if and only if $g \in H^b$. The same is true for a map defined by an analytic function F. More general, this is true for the map $u(x) \mapsto \Pi(F(u(x), x))$ where $F(u, x)$ is a C^b-smooth function of u and x, which is δ-analytic in u with some x-independent $\delta > 0$. $\square$

Given a C^1-smooth (or analytic) functional $H : X_a \supset O_a \to \mathbb{R}$, we identify the adjoint space $(X_a)^*$ with X_{-a} and consider the *gradient map*

$$\nabla H : O_a \to X_{-a}, \quad \langle \nabla H(u), v\rangle = H_*(u)v \quad \forall v \in X_a.$$

If the domain O_a belongs to a system of compatible domains O_s $(a \le s \le b)$ and the gradient map ∇H is a C^1-smooth (or analytic) map of order d_H in this system of domains, we write $\operatorname{ord} \nabla H = d_H$. In this case the linearised gradient map $\nabla H(u)_* : X_s \to X_{s-d_H}$ is bounded for any $u \in O_s$ with $a \le s \le b$. Since

$$\langle \nabla H(u)_* v_1, v_2\rangle = d_2 H(u)(v_1, v_2) = \langle \nabla H(u)_* v_2, v_1\rangle,$$

then the map $\nabla H(u)_*$ is symmetric and due to the Corollary from Theorem 1 it defines bounded linear maps $\nabla H(u)_* : X_s \to X_{s-d_H}$ for $s \in [d_H - d, d]$ (assuming that $2d \ge d_H$).

1.3 Differential forms

For $d \ge 0$ and a domain O in a Hilbert space X_d from a Hilbert scale $\{X_s\}$ we identify tangent spaces $T_{\mathfrak{x}}O$ with X_d and treat differential k-forms on O as continuous functions $O \times \underbrace{(X_d \times \cdots \times X_d)}_{k} \longrightarrow \mathbb{R}$, which are polylinear and skew-symmetric in the last k arguments (see more in [Ca, La]). We write 1-forms as $a(\mathfrak{x})\, d\mathfrak{x}$, where $a : O \to X_{-d}$ and $a(\mathfrak{x})\, d\mathfrak{x}[\xi] \stackrel{\text{def}}{=} \langle a(\mathfrak{x}), \xi\rangle$ for $\xi \in X_d$. Besides, we write 2-forms as $A(\mathfrak{x})\, d\mathfrak{x} \wedge d\mathfrak{x}$, where

$$A(\mathfrak{x})\, d\mathfrak{x} \wedge d\mathfrak{x}[\xi, \eta] \stackrel{\text{def}}{=} \langle A(\mathfrak{x})\xi, \eta\rangle_0 \quad \text{for} \quad \xi, \eta \in X_d,$$

---

[3] if $b = \infty$, then $a \le s < \infty$. This agreement applies everywhere below.

and $A(\mathfrak{x}) : X_d \to X_{-d}$ is a bounded anti selfadjoint operator, i.e., its adjoint $A(\mathfrak{x})^* : X_d \to X_{-d}$ equals $-A(\mathfrak{x})$.

All the forms we consider below are analytic, where a k-form ω_k on $O \subset X_d$ is called analytic if the corresponding map from O to the linear space of skew-symmetric polylinear functions $\underbrace{(X_d \times \cdots \times X_d)}_{k} \longrightarrow \mathbb{R}$ is analytic.

To define the *differential* of a C^1-smooth k-form ω_k we use the Cartan formula:

$$d\omega_k(\mathfrak{x})[\xi_1, \ldots, \xi_{k+1}] = \sum_{i=1}^{k+1} (-1)^{i-1} \frac{\partial}{\partial \xi_i} \omega_k(\mathfrak{x})[\xi_1, \ldots, \hat{\xi}_i, \ldots, \xi_{k+1}]. \qquad (1.2)$$

Here the vectors $\xi_j \in T_{\mathfrak{x}}O \simeq X_d$ are extended to constant vector fields on O. So the r.h.s. of (1.2) is well-defined and the commutator-terms from the r.h.s. of the classical Cartan formula (see e.g. [Go, La]) vanish. This definition well agree with the finite-dimensional situation, as states the following obvious lemma:

Lemma 1. Let ω_k be a k-form on a domain $O \subset X_d$, L be a finite-dimensional affine subspace of X_d and $L^O = L \cap O$. Then $d\omega_k |_{L^O} = d(\omega |_{L^O})$.

Proof. Both forms are given by the same formula (1.2). $\square$

Example 1.2. 1) The differential of a C^1-function f on O (=a zero-form) equals $df = \nabla f(\mathfrak{x}) d\mathfrak{x}$. 2) The differential of a 1-form $a(\mathfrak{x}) d\mathfrak{x}$, $a : O \to X_{-d}$, equals $d(a(\mathfrak{x}) d\mathfrak{x}) = (a(\mathfrak{x})_* - a(\mathfrak{x})^*) d\mathfrak{x} \wedge d\mathfrak{x}$. Indeed, the operator $A(\mathfrak{x}) = a(\mathfrak{x})_* - a(\mathfrak{x})^* : X_d \to X_{-d}$ is anti selfadjoint and $d(a(\mathfrak{x})d\mathfrak{x})[\xi, \eta] = \langle a(\mathfrak{x})_*\xi, \eta \rangle - \langle a(\mathfrak{x})_*\eta, \xi \rangle = \langle A(\mathfrak{x})\xi, \eta \rangle$.

In the sequel we shall also work with k-forms in sub-domains of the direct products Z_d,

$$Z_d = X \times Y_d, \quad Z_d \ni z = (x, y),$$

where X is a finite-dimensional Euclidean space and Y_d is a space from a Hilbert scale $\{Y_s\}$.[4] We write linear operators $\mathfrak{A}$ in Z_d in the block-form,

$$\mathfrak{A} = \begin{pmatrix} \mathfrak{A}_{XX} & \mathfrak{A}_{XY} \\ \mathfrak{A}_{YX} & \mathfrak{A}_{YY} \end{pmatrix},$$

where $\mathfrak{A}_{XY} : Y_d \to X$, $\mathfrak{A}_{YX} : X \to Y_d$ and $\mathfrak{A}_{XX} : X \to X$, $\mathfrak{A}_{YY} : Y_d \to Y_d$ are bounded linear operators. The operator $\mathfrak{A}$ is anti selfadjoint (with respect to the scalar product in $X \times Y_0$) if $\mathfrak{A}_{XY} = -\mathfrak{A}_{YX}^*$ and $\mathfrak{A}_{XX}, \mathfrak{A}_{YY}$ are anti selfadjoint operators. Accordingly we write the 2-form $\mathfrak{A} dz \wedge dz$ as

$$\mathfrak{A}(z) dz \wedge dz = \mathfrak{A}_{XX}(x,y) dx \wedge dx + \mathfrak{A}_{XY}(x,y) dy \wedge dx +$$
$$+ \mathfrak{A}_{YX}(x,y) dx \wedge dy + \mathfrak{A}_{YY}(x,y) dy \wedge dy.$$

---

[4] Obviously, the spaces $\{Z_s\}$ also form a Hilbert scale.

We note that in our notations

$$\mathfrak{A}_{YX}(x,y)\, dx \wedge dy[(\delta x_1, \delta y_1),\, (\delta x_1, \delta y_2)] = \langle \mathfrak{A}_{YX}\delta x_1, \delta y_2\rangle_Y = -\langle \delta x_1, \mathfrak{A}_{XY}\delta y_2\rangle.$$

For sub-domains of the manifolds $\mathcal{Y}_d$,

$$\mathcal{Y}_d = \mathbb{R}^n \times \mathbb{T}^n \times Y_d = \{(p,q,y)\}, \qquad (1.3)$$

we use natural versions of the notations given above.

The Poincarè lemma states that "locally" each closed form is exact. The proof is constructive and is well applicable to infinite-dimensional problems (see [Ca, La]). We shall need a version of the lemma for a closed 2-form defined in a neighbourhood $O \subset \mathcal{Y}_d$ of the set $P \times \mathbb{T}^n \times \{0\}$, where P is a sub-domain of $\mathbb{R}^n$, such that fibres of the natural fibration $O \to \mathbb{R}^n \times \mathbb{T}^n$ are convex. Below we state the result, denoting by w points from $\mathbb{R}^n \times \mathbb{T}^n$:

Lemma 2. *If $\omega_2(w,y)$ is a closed 2-form in O and $\omega_2(w,0) = 0$, then $\omega_2 = d\omega_1$, where*

$$\omega_1(w,y)(\delta w, \delta y) = \int_0^1 \omega(w,ty)[(0,y),(\delta w, t\delta y)]\, dt.$$

In particular, if $\omega_2 = A_{WW}(w,y)dw \wedge dw + A_{WY}(w,y)\, dy \wedge dw - A_{WY}^*(w,y)\, dw \wedge dy$, then $\omega_1 = a(w,y)\, dw$, where $a(w,y) = \left(\int_0^1 A_{WY}(w,ty)\, dt\right) y$.

This result follows from its finite-dimensional version (see [AG]) and Lemma 1. Indeed, For any $(w,y) \in O$ and $\xi_1, \xi_2 \in T_{(w,y)}\mathcal{Y}_d \simeq \mathbb{R}^{2n} \times Y_d = Z_d$ we denote by Q a sufficiently small neighbourhood of (w,y) in $\mathcal{Y}_d$ and treat Q as a domain in Z_d. Now we take for L the affine 3-space through (w,y) in the directions $(0,y), \xi_1, \xi_2$ and get that $d\omega_1(w,y)[\xi_1, \xi_2] = \omega_2(w,y)[\xi_1, \xi_2]$.

2 Symplectic structures and Hamiltonian equations

2.1 Basic definitions

In a domain $O_d \in X_d$ with $d \geq 0$ let us take a closed 2-form $\alpha_2 = \bar{J}(\mathfrak{x})\, d\mathfrak{x} \wedge d\mathfrak{x}$ such that the operator $\bar{J}$ analytically depends on $\mathfrak{x} \in O_d$ and defines a linear isomorphism $\bar{J}(\mathfrak{x}) : X_d \xrightarrow{\sim} X_{d+d_J}$, $d_J \geq 0$. Since the operator is anti selfadjoint in X_0, then by the Corollary from Theorem 1 it defines an anti selfadjoint automorphism of order $-d_J$ of the scale $\{X_s\}$ for $s \in [-d - d_J, d]$. The form α_2 supplies O_d with a symplectic structure which corresponds to any C^1-smooth function h on O_d the Hamiltonian vector field V_h defined by the usual (see [A1]) relation: $\alpha_2[V_h, \xi] = -dh(\xi)$ for all $\xi \in TO_d$. For any $\mathfrak{x} \in O_d$ we have $\langle \bar{J}(\mathfrak{x})V_h(\mathfrak{x}), \xi\rangle = -\langle \nabla h(\mathfrak{x}), \xi\rangle$ for each $\xi \in X_d$. Thus,

$$V_h(\mathfrak{x}) = J(\mathfrak{x})\nabla h(\mathfrak{x}), \quad \text{where} \quad J = (-\bar{J})^{-1}.$$

The operator $J(\mathfrak{x})$ is called *an operator of the Poisson structure*. For each $\mathfrak{x}$ it defines an anti selfadjoint automorphism of the scale of order d_J and the operator

$$J(\mathfrak{x}) : X_{s+d_J} \overset{\sim}{\rightleftarrows} X_s, \quad -d - d_J \leq s \leq d, \tag{2.1}$$

analytically depend on $\mathfrak{x} \in O_d$ (see the Corollary to Theorem 1). Since the functional h is C^1-smooth, then the gradient map $\nabla h : O_d \to X_{-d}$ is continuous. Using (2.1) we get that the vector field V_h defines a continuous map $O_d \to X_{-d-d_J}$. Usually we shall impose an additional restriction and assume that the vector field V_h is smoother than that and ord $V_h =: d_1 < 2d + d_J$.

To stress that a domain $O_d \subset X_d$ is given a symplectic structure as above we shall write it as a pair (O_d, α_2). If the form α_2 is analytic on the whole space X_s for each $s \geq s_0$ with some fixed s_0, we shall say that $(\{X_s\}, \alpha_2)$ is a *symplectic Hilbert scale*. A basis $\{\phi_j(\mathfrak{x}) \mid j \in \mathbb{Z}_n\}$ of a tangent space $T_{\mathfrak{x}} O_d = X_d$ is called *symplectic* if

$$\alpha_2[\phi_j(\mathfrak{x}), \phi_k(\mathfrak{x})] = \nu_j(\mathfrak{x})\delta_{j,-k} \tag{2.2}$$

for any $j \in \mathbb{N}$ and each $k \in \mathbb{Z}_0$, with some positive real numbers $\nu_j(\mathfrak{x}), j \in \mathbb{N}$.

For any C^1-smooth function h on $O_d \times \mathbb{R}$ a Hamiltonian equation with the hamiltonian $h(\mathfrak{x}, t)$ is the equation

$$\dot{\mathfrak{x}}(t) = J(\mathfrak{x})\nabla h(\mathfrak{x}, t) =: V_h(\mathfrak{x}, t). \tag{2.3}$$

If ord $V_h = 0$ and the vector field V_h is C^1-smooth and Lipschitz in O_d, then the initial-value problem for the equation (2.3) is well-posed: for any given initial condition $\mathfrak{x}(0) \in O_d$ it has a unique solution defined while it stays in O_d.

A partial differential equation, supplemented by appropriate boundary conditions, is called a *Hamiltonian PDE* if under a suitable choice of a symplectic Hilbert scale $(\{X_s\}, \alpha_2)$, a domain $O_d \subset X_d$ and a hamiltonian h, the equation can be written in the form (2.3). In this case the vector field V_h is unbounded, ord $V_h = d_1 > 0$:

$$V_h : O_d \times \mathbb{R} \to X_{d-d_1}. \tag{2.4}$$

Usually the domain O_d belongs to a system of compatible domains O_s, $s \geq d_0$, and the map V_h is analytic of order d_1 for $s \geq d_0$.

For a vector field V_h as in (2.4) with $d_1 > 0$, different classes of solutions for (2.3) can be considered. We choose the following definition: a continuous curve $\mathfrak{x} : [0, T] \to O_d$ is a *solution of (2.3) in a space X_d* if it defines a C^1-smooth map $[0, T] \to X_{d-d_1}$ and both parts of (2.3) coincide as curves in X_{d-d_1}. A solution $\mathfrak{x}(t)$ is called *smooth* if it defines a smooth curve in each space X_l.

If a solution $\mathfrak{x}(t)$, $t \geq \tau$, of (2.3) with $\mathfrak{x}(\tau) = \mathfrak{x}_\tau$ exists and is unique, we write $\mathfrak{x}(t) = S^t_\tau \mathfrak{x}_\tau$, or $\mathfrak{x}(t) = S^{t-\tau} \mathfrak{x}_\tau$ if the equation is autonomous. The operators S^t_τ and S^τ are called *flow-maps* of the equation.

For an equation (2.3) with $d_1 > 0$ there is no general existence theorem for a solution of the corresponding initial-value problem which would guarantee existence of the flow-maps. To prove the existence is an art we do not touch here.

Example 2.1 (semilinear equation). If (2.3) is a *semilinear equation*, i.e., $V_h = A + V^0$, where A is an unbounded linear operator with a discrete imaginary spectrum and the nonlinearity V^0 is Lipschitz on bounded subsets of X_d, then solutions of the equation with prescribed initial conditions are well defined till they stay in a bounded part of X_d, see [Paz]. Some important Hamiltonian PDEs are semilinear. For example, the nonlinear Schrödinger equation:

$$\dot{u}(t, x) = i(\Delta u + f(|u|^2)u), \quad x \in S^1,$$

where f is a smooth real-valued function (see [K, Paz]). Still, the semi-linearity assumption is very restrictive since it fails for many important Hamiltonian PDEs (e.g., for the KdV).

Our main concern are *quasilinear Hamiltonian equations*, i.e., equations (2.3) with $V_h = A + V^0$, where A is a liner operator and $\operatorname{ord} A > \operatorname{ord} V^0$. Possibly $\operatorname{ord} V^0 > 0$ i.e., the equation may be non-semilinear.

Let $Q \subset O_d$ be a sub-domain such that the flow-maps maps $S_\tau^t : Q \to O_d$ are well-defined and are C^1-smooth for $\tau \leq t \leq T$, where $-\infty \leq \tau < T \leq \infty$. Then differentiating a solution $\mathfrak{x}(t)$ of (2.3) in the initial condition we get that the curve $\zeta(t) := S_\tau^t(\mathfrak{x}(\tau))_* \zeta$ satisfies the linearised equation

$$\dot{\zeta}(t) = V_h(\mathfrak{x}(t), t)_* \zeta(t), \quad \zeta(\tau) = \zeta. \qquad (2.5)$$

The assumption that the map S_τ^t is C^1-smooth in a sub-domain is very restrictive since to check the smoothness of flow-maps for many important equations (even for the KdV!) is a nontrivial task. To get rid of it we give the following

Definition 1. Let $\mathfrak{x}(t)$, $t \in \mathbb{R}$, be a solution for equation (2.3). If for each $\zeta \in X_d$ and each θ the linearised equation $\dot{\zeta}(t) = V_h(\mathfrak{x}(t), t)_* \zeta(t)$, $\zeta(\theta) = \zeta$, has a unique solution $\zeta(t) \in X_d$ defined for all t and such that $\|\zeta(t)\|_d \leq C\|\zeta\|_d$ uniformly in θ, t from a compact segment, then we write $\zeta(t) = S_{\theta*}^t(\mathfrak{x})\zeta$ and say that *flow* $\{S_{\theta*}^t(\mathfrak{x})\}$ *of the linearised equation (2.5) is well defined* in X_d.

The property described in Definition 1 characterises the flow only in the "infinitesimal vicinity" of a solution of (2.3). It suits well our goal to study special families of solutions rather than the whole flow of the equation. If the flow-maps S_τ^t are C^1-smooth, then $S_\tau^t(\mathfrak{x})_* = S_{\tau*}^t(\mathfrak{x})$, but the map in the r.h.s. of this relation can be well defined while the map in the l.h.s. is not.

Example 2.2 (Equations of the Korteweg–de Vries type). Let us take for $\{X_s\}$ the scale of Sobolev spaces H_0^s as in Example 1.1. We define a Poisson structure by means of the operator $J = \frac{\partial}{\partial x}$, so $d_J = 1$ and $-\bar{J}$

is the operator $(\partial/\partial x)^{-1}$ of integrating with zero mean-value. We get the symplectic Hilbert scale $(\{H_0^s\}, -(\partial/\partial x)^{-1}du \wedge du)$. We stick to the discrete scale $\{s \in \mathbb{Z}\}$: it is sufficient since the orders of all involved operators are integer. The trigonometric basis $\{\varphi_j \mid j \in \mathbb{Z}_0\}$ (see (1.1)) is symplectic since for $j \geq 1$ and any k we have:

$$\alpha_2[\varphi_j, \varphi_k] = \langle \bar{J}\pi^{-1/2}\cos jx, \varphi_k(x)\rangle = j^{-1}\langle -\pi^{-1/2}\sin jx, \varphi_k(x)\rangle = j^{-1}\delta_{j,-k}.$$

For a hamiltonian h we take $h(u) = \int_0^{2\pi}(-\frac{1}{8}u'(x)^2 + f(u))\,dx$ with some analytic function $f(u)$. Then $\nabla h(u) = \frac{1}{4}u'' + f'(u)$ [5] and $V_h(u) = \frac{1}{4}u''' + \frac{\partial}{\partial x}f'(u)$. Thus the Hamiltonian equation takes the form

$$\dot{u}(t,x) = \frac{1}{4}u''' + \frac{\partial}{\partial x}f'(u) \qquad (2.6)$$

(for $f(u) = \frac{1}{4}u^3$ we get the KdV equation). The maps $H_0^s \to \mathbb{R}$, $u(x) \mapsto \int f(u)\,dx$ and $H_0^s \to H^s$, $u(x) \mapsto f'(u(x))$, with $s \geq 1$ are analytic (see Example 1.1). So the map $H_0^s \ni u \mapsto V_h(u) \in H_0^{s-3}$ is analytic for $s \geq 1$. That is, the vector field V_h is analytic of order 3 for $s \geq 3$.

Being supplemented by an initial condition $u(0,x) = u_0(x) \in H_0^s$ with $s \geq 3$, equation (2.6) has a unique solution in H_0^s. This solution exists for $t < T(\|u_0\|_s)$ (T is a continuous positive function) and the flow-maps S^t, $t < T$, are C^1-smooth. This is a non-trivial result, see e.g. [Kat1]. On the contrary, if $u(t,x)$ is a smooth solution of (2.6), then the linearised equation

$$\dot{v} = \frac{1}{4}v''' + \frac{\partial}{\partial x}(f''(u)v), \quad v(0,x) = v_0 \in H_0^s, \qquad (2.7)$$

has a unique solution in H_0^s with any $s \geq 0$ by *trivial* arguments, see [KK, Paz].

We shall often work with equations in a sub-domain O_d of the manifold $\mathcal{Y}_d$ ($d \geq 0$) as in (1.3), given a symplectic structure by means of a 2-form $(dp \wedge dq) \oplus (\bar{\Upsilon}(y)dy \wedge dy)$, where $dp \wedge dq$ is the classical symplectic form on $\mathbb{R}^n \times \mathbb{T}^n$ and $\bar{\Upsilon}(y)dy \wedge dy$ is a closed 2-form in a domain in Y_d. This symplectic structure corresponds to a C^1-smooth function $H(p,q,y)$ the following Hamiltonian system:

$$\dot{p} = -\nabla_q H, \quad \dot{q} = \nabla_p H, \quad \dot{y} = \Upsilon \nabla_y H.$$

Solutions of these equations are defined in the same way as solutions of (2.3).

2.2   Symplectic transformations

Let $\{X_s\}$, $\{Y_s\}$ be two Hilbert scales and $d, \bar{d} \geq 0$. Let $O \subset X_d$ and $Q \subset Y_{\bar{d}}$ be two domains given symplectic structures by 2-forms $\alpha_2 = \bar{J}(\mathfrak{x})d\mathfrak{x} \wedge d\mathfrak{x}$ and

---

[5] since $dh(u)v = \int -\frac{1}{4}u'(x)v'(x) + f'(u(x))v(x)\,dx = \langle\frac{1}{4}u''(x) + f'(u(x)), v(x)\rangle_{L_2}$.

$\beta_2 = \bar{\Upsilon}(y)dy \wedge dy$ as in Section 2.1. A C^1-smooth map $\Phi : Q \to O$ is called *symplectic* if $\Phi^*\alpha_2 = \beta_2$. That is, if for any $y \in Q$ with $\Phi(y) = \mathfrak{x} \in O$ we have $\langle \bar{J}(\mathfrak{x})\Phi_*(y)\xi, \Phi_*(y)\eta \rangle_{X_0} = \langle \bar{\Upsilon}(y)\xi, \eta \rangle_{Y_0}$ for all $\xi, \eta \in Y_{\bar{d}}$, or

$$\Phi^*(y) \circ \bar{J}(\mathfrak{x}) \circ \Phi_*(y) = \bar{\Upsilon}(y). \tag{2.8}$$

A symplectic map Φ is an immersion since by (2.8) its tangent maps are embeddings. If a symplectic map Φ is such that the tangent maps $\Phi_*(y)$ define isomorphisms of the spaces $Y_{\bar{d}}$ and X_d, then Φ is called a *symplectomorphism*.

We shall need an obvious version of the definitions above for the case when O^c and Q^c are *complex* domains in $X_{\bar{d}}^c$ and $Y_{\bar{d}}^c$ respectively and the forms α_2, β_2 have constant coefficients: an analytic map $\Phi_1 : (Q^c, \alpha_2) \to (O^c, \beta_2)$ is symplectic if $\langle \bar{J}\Phi_1(y)_*\xi, \Phi_1(y)_*\eta \rangle_{X_0} \equiv \langle \bar{\Upsilon}\xi, \eta \rangle_{Y_0}$. In particular, an analytic symplectomorphism $\Phi : Q \to O$ analytically extends to a symplectomorphism of sufficiently small complex neighbourhoods of Q and O (the forms α_2, β_2 are assumed to be constant-coefficient).

From now on for the sake of simplicity we restrict ourselves to the case we need below:

$$d = \bar{d} \geq 0, \quad \operatorname{ord} \bar{J}(\mathfrak{x}) = \operatorname{ord} \bar{\Upsilon}(y) = -d_J \quad \forall \mathfrak{x}, y.$$

Proposition 1. *Let us assume that $\bar{J}(\mathfrak{x}) = \bar{J}$ and $\bar{\Upsilon}(y) = \bar{\Upsilon}$ are constants isomorphisms of the corresponding scales of order $-d_J$. Let $\Phi : (Q^c, \beta_2) \to (O^c, \alpha_2)$ be a symplectomorphism such that $\|\Phi(y)_*\|_{d,d}, \|(\Phi(y)_*)^{-1}\|_{d,d} \leq C$ for every $y \in Q^c$. Then $\|\Phi(y)_*\|_{\theta,\theta}, \|(\Phi(y)_*)^{-1}\|_{\theta,\theta} \leq C_1$ for every $y \in Q^c$ and every $\theta \in [-d - d_J, d]$. If Φ is an analytic symplectomorphism, then these maps are analytic in $y \in Q^c$.*

Proof. By (2.8) we have $\Phi^* = -\bar{\Upsilon} \circ \Phi_* \circ J$. So $\|\Phi(y)^*\|_{d+d_J, d+d_J} \leq C'$ for every y. Hence, $\|\Phi(y)_*\|_{-d-d_J, -d-d_J} \leq C'$ and the estimate for $\|\Phi_*\|_{\theta,\theta}$ follows by interpolation. The estimates for Φ_*^{-1} follow from the identity $(\Phi^*)^{-1} = -\bar{J} \circ \Phi_* \circ \Upsilon$ since $\Phi_*^{-1} = \left((\Phi^*)^{-1}\right)^* = (-\bar{J} \circ \Phi_* \circ \Upsilon)^*$ is a zero-order morphism for $s \in [-d - d_J, d]$.

In the analytic case the maps are analytic in y by the criterion of analyticity. $\square$

As in the finite-dimensional case, symplectic maps transform Hamiltonian equations to Hamiltonian. Let $\Phi : Q \to O^c$ be an analytic symplectic map such that

$$\Phi(y)_* : Y_s \to X_s \text{ is a linear map, analytic in } y \in Q, \text{ for any } |s| \leq d. \tag{2.9}$$

We note that if $\bar{J}(x) = \bar{J}$ and $\bar{\Upsilon}(y) = \bar{\Upsilon}$ are constant isomorphisms of zero order, then the assumption (2.9) is satisfied due to Proposition 1.

Theorem 2. *Let domains $O \subset X_d$ and $Q \subset Y_d$, $d \geq 0$, be given symplectic structures by analytic 2-forms α_2, β_2 as above with $\operatorname{ord} \bar{J} = \operatorname{ord} \bar{\Upsilon} = -d_J$.*

Let the vector field $V_h = J\nabla h$ of equation (2.3) defines a C^1-smooth map $V_h : O \times \mathbb{R} \to X_{d-d_1}$ of order $d_1 \leq 2d$ and let $\Phi : Q \to O$ be an analytic symplectic map satisfying (2.9), such that the vector field V_h in O is tangent to $\Phi(Q)$ in the following sense:

$$V_h(\Phi(y)) = \Phi(y)_*\xi \quad \text{for any } y \in Q \text{ with an appropriate } \xi \in Y_{d-d_1}.$$
$$(2.10)$$

$$\dot{y} = \Upsilon(y)\nabla_y H(y,t), \quad H = h \circ \Phi, \quad \Upsilon = (-\tilde{\Upsilon})^{-1}, \qquad (2.11)$$

to solutions of (2.3).

We note right away that the assumption (2.10) becomes empty if Φ is a symplectomorphism (to be more specific, now (2.10) follows from (2.9) since $d - d_1 \geq -d$).

Proof. Let $y(t)$ be a solution of (2.11). By (2.9) the curve $\mathfrak{x}(t) = \Phi(y(t))$ is C^1-smooth in Y_{d-d_1} and is continuous in Y_d. It remains to check that it satisfies (2.3). Since $\dot{\mathfrak{x}} = \Phi(y)_*\dot{y}$ and $\nabla_y H = \Phi(y)^*\nabla_{\mathfrak{x}} h$, then

$$\dot{\mathfrak{x}} = \Phi(y)_*\Upsilon(y)\Phi(y)^*\nabla_{\mathfrak{x}} h = -\Phi(y)_*\Upsilon(y)\Phi(y)^*\bar{J}(\mathfrak{x})V_h(\mathfrak{x}).$$

By (2.10), $V_h(\mathfrak{x}) = \Phi(y)_*\xi$. So the r.h.s is $-\Phi(y)_*\Upsilon(y)\Phi(y)^*\bar{J}(\mathfrak{x})\Phi(y)_*\xi$. By (2.8) it equals $-\Phi(y)_*\Upsilon(\mathfrak{x})\tilde{\Upsilon}(\mathfrak{x})\xi = \Phi(y)_*\xi = V_h(\mathfrak{x})$. Thus, $\mathfrak{x}(t)$ satisfies the equation (2.3). □

To apply Theorem 2 we have to be able to construct sufficient amount of symplectic transformations. An important way to construct symplectomorphisms of sub-domains of $(O \subset X_d, \alpha_2)$ is to get them as flow-maps S_t^τ of an additional nonautonomous Hamiltonian equation $\dot{\mathfrak{x}} = J(\mathfrak{x})\nabla_{\mathfrak{x}} f(t,\mathfrak{x}) = V_f(t,\mathfrak{x})$, where the hamiltonian f is such that the vector field V_f is Lipschitz, see [K, KK].

Let O be a domain in a symplectic space $(X_d, \alpha_2 = \bar{J}(\mathfrak{x}) \, d\mathfrak{x} \wedge d\mathfrak{x})$.

Definition 2. Let C^1-smooth functions H_1, H_2 on a domain $O \subset X_d$ define continuous gradient maps of orders $d_1, d_2 \leq 2d$ such that $d_1 + d_2 + d_J \leq 2d$. Then the Poisson bracket $\{H_1, H_2\}$ of H_1, H_2 is the continuous on O function $\{H_1, H_2\}(\mathfrak{x}) = \langle J(\mathfrak{x})\nabla H_1(\mathfrak{x}), \nabla H_2(\mathfrak{x})\rangle$.

The scalar product $\langle J(\mathfrak{x})\nabla H_1, \nabla H_2\rangle$ is well-defined and is continuous in $\mathfrak{x}$. The Poisson bracket is skew-symmetric, $\{H_1, H_2\} = -\{H_2, H_1\}$. In particular, $\{H, H\} \equiv 0$ (if ord $\nabla H \leq d - d_J/2$).

2.3 Darboux lemmas

The classical Darboux lemma states that locally near a point any closed non-degenerate 2-form in $\mathbb{R}^{2n}$ can be written as $dp \wedge dq$. This result has several versions which put a closed non-degenerate 2-form on a manifold to different normal forms in the vicinity of a closed set (for the classical lemma the set is a point), see [AG]. Some of these results admit direct infinite-dimensional reformulations which can be proven by the same arguments due to Moser–Weinstein. In this section we present a version of the Darboux lemma which will be used later on.

Let $\mathcal{Y}_d = \mathbb{R}^n \times \mathbb{T}^n \times Y_d$ and W be its subset of the form $W = P \times \mathbb{T}^n \times \{0\}$, where P is a bounded sub-domain of $\mathbb{R}^n$. By $O, O_1, \ldots$ we denote δ-neighbourhoods of W in $\mathcal{Y}_d$ with different $\delta > 0$ and suppose that in a neighbourhood O we are given two closed analytic 2-forms ω_0 and ω_1, both of them of the form $\omega_j = \bar{J}^j(\mathfrak{y}) \, d\mathfrak{y} \wedge d\mathfrak{y}$, where $\mathfrak{y} = (p, q, y)$ and $d\mathfrak{y} = (dp, dq, dy)$. We assume that

(i) $\omega_0 = \omega_1$ in $TO \,|_W$,
(ii) for all $t \in [0, 1]$ and all $\mathfrak{y} \in O$ the map $\bar{J}^t(\mathfrak{y}) = (1 - t)\bar{J}^0 + t\bar{J}^1$ defines an isomorphism $\bar{J}^t : Z_d \xrightarrow{\sim} Z_{d+d_J}$ with some fixed $d_J \geq 0$, where $Z_d = \mathbb{R}^n \times \mathbb{R}^n \times Y_d$.

By (ii), the map $J^t = (-\bar{J}^t)^{-1} : Z_{d+d_J} \xrightarrow{\sim} Z_d$ is well defined and analytically depends on $\mathfrak{y}$. By Poincaré's lemma (see Lemma 2 above), the form $\omega_1 - \omega_0$ equals $d\alpha$ for some analytic one-form $\alpha = a(\mathfrak{y}) \, d\mathfrak{y}$ such that $a(p, q, y) = O(\|y\|_d^2)$. Let us also assume that

(iii) the map $O_1 \to Z_{d+d_J}$, $\mathfrak{y} \mapsto a$, is Lipschitz analytic in a neighbourhood O_1 of W.

Lemma 3 (Moser–Weinstein). *Under the assumptions i)-iii) there exists a neighbourhood O_2 and an analytic diffeomorphism $\varphi : O_2 \to O$ such that $\varphi \,|_W = \mathrm{id}$, $\varphi_* \,|_W = \mathrm{id}$ and $\varphi^* \omega_1 = \omega_0$. Moreover, φ equals to a time-one flow-map S_0^1, corresponding to the non-autonomous equation $\dot{\mathfrak{y}} = J^t a(\mathfrak{y}) =: V(t, \mathfrak{y})$.*

Proof. For $0 \leq t \leq 1$ let us consider the 2-forms $\omega_t = (1 - t)\omega_0 + t\omega_1 = \bar{J}^t \, d\mathfrak{y} \wedge d\mathfrak{y}$. These forms are closed as well as the forms ω_0, ω_1 and are non-degenerate in a neighbourhood O_3 since $\omega_t = \omega_0 = \omega_1$ on W by i). Now we denote by φ^t the flow-maps S_0^t of equation $\dot{\mathfrak{y}} = V(t, \mathfrak{y})$; so $\varphi^0 = \mathrm{id}$, $\varphi^1 = \varphi$ and $(\varphi^1 - \mathrm{id})(p, q, y) = O(\|y\|_d^2)$. The lemma will be proven if we check that $(\varphi^t)^* \omega_t = \mathrm{const}$. Because Cartan's identity (see e.g. [GS] and see [KK] for the infinite-dimensional case) we have to prove that $\frac{\partial \omega_t}{\partial t} + d(V \rfloor \omega_t) = 0$. Since $\partial \omega_t / \partial t = \omega_1 - \omega_0 = d\alpha$, then it remains to check that $\alpha + V \rfloor \omega_t \equiv 0$. But $V \rfloor \omega_t = V \rfloor \bar{J}^t \, d\mathfrak{y} \wedge d\mathfrak{y} = (\bar{J}^t V) d\mathfrak{y}$. So $\alpha + V \rfloor \omega_t = (a + \bar{J}^t V) \, d\mathfrak{y} = (a + \bar{J}^t J^t a) \, d\mathfrak{y} = 0$ and the lemma is proven. $\square$

Appendix. Time-quasiperiodic solutions

Our main goal is to study time-quasiperiodic solutions $\mathfrak{x}(t)$ of some Hamiltonian equations (2.3). Here we recall corresponding basic definitions.

Definition. *A C^1-curve $\gamma : \mathbb{R} \to X$ in a Banach space or a manifold X is called quasiperiodic (QP) with $\leq n$ frequencies if there exists a C^1-smooth map $\Gamma : \mathbb{T}^n \to X$, a vector $\omega \in \mathbb{R}^n$ and a point $q_0 \in \mathbb{T}^n$ such that*

$$\gamma(t) \equiv \Gamma(q_0 + \omega t). \tag{A.1}$$

The vector ω is called the frequency vector and q_0 is called the phase. The minimal n such that $\gamma(t)$ admits a representation (A.1) is called the number of independent frequencies; corresponding numbers $\omega_1, \ldots, \omega_n$ are called the basic frequencies and the numbers $2\pi/\omega_1, \ldots, 2\pi/\omega_n$ – the basic periods.

Remark. We note that the vector ω formed by the basic frequencies is defined only up to an unimodular transformation L since the curve $\gamma(t)$ can be also written as $\gamma(t) = \Gamma_L(Lq_0 + L\omega t)$, where $\Gamma_L(q) = \Gamma(L^{-1}q)$. What is uniquely defined, is the $\mathbb{Z}$-module $\mathbb{Z}\omega_1 + \mathbb{Z}\omega_2 + \cdots + \mathbb{Z}\omega_n \subset \mathbb{R}$. We shall usually ignore this subtlety.

The closure $\overline{\gamma(\mathbb{R})}$ is called the *hull* of γ. If components of ω are rationally independent, then the hull equals $\Gamma(\mathbb{T}^n)$.

We call a solution $\mathfrak{x}$ of (2.3) a (time-) quasiperiodic, or a QP solution, if the curve $\mathfrak{x} : \mathbb{R} \to X_d$ is QP, and call it analytic quasiperiodic if the corresponding map Γ is analytic of maximal rank. The hull $\Gamma(\mathbb{T}^n)$ of an analytic QP solution (A.1) with n basic frequencies is an invariant analytic n-torus of the equation. This torus is an analytic submanifold of X if the map Γ is an immersion.

3   Lax-integrable Hamiltonian equations and their integrable subsystems

We consider a symplectic Sobolev scale $(\{Z_s\}, \alpha_2)$, $\alpha_2 = \bar{J} \, dz \wedge dz$, as above. The operator $\bar{J}$ defines an anti selfadjoint automorphism of the scale of a negative order $-d_J \leq 0$. To a hamiltonian $\mathcal{H} = \frac{1}{2}\langle Az, z \rangle + H(z)$, where A is a selfadjoint morphism of the scale of order d_A, the symplectic structure corresponds the Hamiltonian equation

$$\dot{u} = J\nabla\mathcal{H}(u) = J(Au + \nabla H(u)) =: V_{\mathcal{H}}(u), \quad J = (-\bar{J})^{-1}. \tag{3.1}$$

We assume that the hamiltonian $\mathcal{H}$ is analytic quasilinear, that is, the functional H is analytic on a domain $O_d \subset Z_d$, $d \geq d_A/2$, and defines an analytic gradient map of order $d_H < d_A$, $\nabla H : O_d \to Z_{d-d_H}$. By this assumption and the Corollary to Theorem 1, for any $u \in O_d$ the linear map $\nabla H(u)_*$ defines a morphism of the scale $\{Z_s\}$ of order d_H for $s \in [-d + d_H, d]$.

We shall study equation (3.1) in a sufficiently smooth space Z_d taking any d such that $d_1 := \mathrm{ord}\, V_{\mathcal{H}} = d_A + d_J \leq 2d + d_J$. We *do not* assume that the flow maps of the equation are defined on the whole domain O_d (i.e., we do not assume that the equation can be solved for any initial condition $u(0) \in O_d$).

3.1 Examples of Hamiltonian PDEs

Quasilinear Hamiltonian PDEs with analytic coefficients have the form (3.1), where usually O_d equals to the whole space Z_d and the gradient map ∇H is analytic of some order d_H for all sufficiently smooth spaces Z_d (i.e., it is an analytic map of order d_H for $d \geq d_0$ with some fixed d_0). The following examples and their perturbations will be the main through our work:

Example 3.1 (KdV equation, cf. Example 2.2). Let us take for a scale $\{Z_s \mid s \in \mathbb{Z}\}$ the scale $\{H_0^s(S^1; \mathbb{R})\}$ of 2π-periodic Sobolev functions with zero mean value, defined in Example 1.1. We choose $J = \partial/\partial x$, $A = \frac{1}{4}\partial^2/\partial x^2$ and $H(u) = \frac{1}{4}\int u^3\, dx$, so $\mathcal{H}(u) = \int \left(-\frac{1}{8}u'^2 + \frac{1}{4}u^3\right) dx$. The equation (3.1) takes the form:

$$\dot{u} = \frac{1}{4}\frac{\partial}{\partial x}(u_{xx} + 3u^2). \qquad \text{(KdV)}$$

It is considered under zero mean-value periodic boundary conditions:

$$u(t,x) \equiv u(t, x + 2\pi), \qquad \int_0^{2\pi} u(t,x)\, dx \equiv 0,$$

which are satisfied automatically since we are looking for solutions in a space H_0^s. The gradient map $Z_d \to Z_d$, $u \mapsto \nabla H = \frac{3}{4}u^2$, is analytic of order $d_H = 0$ for $d \geq 1$ (see in Example 2.2). Now we have $\mathrm{ord}\, A = d_A = 2$ and $\mathrm{ord}\, J = d_J = 1$. $\square$

Example 3.2 (higher KdV equations). The KdV equation in the previous example is an equation from an infinite hierarchy of Hamiltonian equations, called the KdV-hierarchy [DMN, McT, ZM]. The l-th equation from the hierarchy can be written as an equation (3.1) in the same symplectic Hilbert scale $(\{H_0^s\}, \langle J\, du, du\rangle)$. It has a hamiltonian $\mathcal{H}_l$ of the form

$$\mathcal{H}_l(u) =$$
$$K_l \int_0^{2\pi} \left((-1)^l u^{(l)^2} + \langle\ \text{higher order terms with}\ \leq l - 1\ \text{derivatives}\rangle\right) dx,$$

where K_l is a non-zero constant ($\mathcal{H}_1$ is just the KdV-hamiltonian). In particular, the hamiltonian $\mathcal{H}_2$ has the form $\mathcal{H}_2 = \frac{1}{8}\int(u_{xx}^2 - 5u^2 u_{xx} - 5u^4)\, dx$ and the corresponding Hamiltonian equation is the fifth order partial differential equation:

$$\dot{u} = \frac{1}{4}u^{(5)} - \frac{1}{4}\frac{\partial}{\partial x}(5u_x^2 + 5uu_{xx} + 10u^3).$$

The gradient map of the non-quadratic part of this hamiltonian defines in the Sobolev scale $\{H_0^s\}$ an analytic map of order $d_H = 2$ for any $s \geq 2$. The order d_A of the linear part equals 4 and $d_J = 1$. $\square$

Example 3.3 (Sine-Gordon equation). The Sine-Gordon (SG) equation on the circle,

$$\ddot{u} = u_{xx}(t, x) - \sin u(t, x), \qquad x \in S = \mathbb{R}/2\pi\mathbb{Z}, \qquad \text{(SG)}$$

corresponds to a bounded nonlinearity. For any $u_0 \in H^s(S)$ and $u_1 \in H^{s-1}(S)$ this equation has a unique solution $u(t, x) \in C(\mathbb{R}, H^s) \cap C^1(\mathbb{R}, H^{s-1})$ such that $u(0, x) = u_0$ and $\dot{u}(0, x) = u_1$. This is almost obvious, see [Paz].

The equation (SG) can be written in a Hamiltonian form in many different ways. The most strightforward way is to write (SG) as

$$\dot{u} = -v, \qquad \dot{v} = -u_{xx} + \sin u(t, x). \qquad (3.2)$$

To see that these equations are Hamiltonian, we take the symplectic scale $(\{Z_s = H^s(S) \times H^s(S)\}, \alpha_2 = \langle \bar{J} d\xi, d\xi \rangle)$, where $\xi = (u, v) \in Z_s$ and $J(u, v) = (-v, u)$ (so $\bar{J} = J$). For a hamiltonian $\mathcal{H}$ we choose $\mathcal{H} = \frac{1}{2}\langle A\xi, \xi \rangle + H(\xi)$, where $A(u, v) = (-u_{xx}, v)$ and $H(u, v) = -\int \cos u(x)\, dx$. Then $\nabla H(u, v) = (\sin u, 0)$ and the Hamiltonian equation $\dot{\xi} = J\nabla\mathcal{H} = J(A\xi + \nabla H(\xi))$ coinsides with (3.2).

The Hamiltonian form (3.2) is traditional (cf. [McK]) and is convenient to study explicit ("finite-gap") solutions of (SG), but not to carry out detailed analysis of the equation since the linear operator A as above defines a self-adjoint morphism of the scale $\{Z_s\}$ of order two, which is not an order-two automorphism (the invererse map A^{-1} defines a morphism of order 0, not -2).

To get a Hamiltonian form of the SG equation, convenient for its analysis, we denote $w = A^{-1/2}v$, where $A = -\partial^2/\partial x^2 + 1$, and write (SG) as

$$\dot{u} = -\sqrt{A}\,w, \quad \dot{w} = \sqrt{A}\left(u + A^{-1}(\sin u - u)\right). \qquad (3.3)$$

The equations (3.3) have more symmetric form than (3.2). They turn out to be Hamiltonian in the shifted Sobolev scale $\{Z_s = H^{s+1}(S) \times H^{s+1}(S)\}$, see [K, BiK1] and [KK].

Even periodic boundary conditions. If $u(t, x)$ is any solution of (SG) such that the initial conditions $(u(0, x), \dot{u}(0, x)) = (u_0(x), u_1(x))$ are even periodic functions, i.e.,

$$u_0(x) \equiv u_0(x + 2\pi) \equiv u_0(-x) \qquad \text{(EP)}$$

and similar with u_1, then $u^-(t, x) = u(t, -x)$ is another 2π-periodic solution for SG with the same initial conditions. Since a solution of the initial-value problem for (SG) is unique, then $u^-(t, x) \equiv u(t, x)$. That is, the space of even periodic functions is invariant under the SG-flow and we can study the equation under the boundary conditions (EP). These conditions clearly imply Neuman boundary conditions on the half-period:

$$(u'_{0x}, u'_{1x})(0) = (u'_{0x}, u'_{1x})(\pi) = (0, 0). \qquad \text{(N)}$$

The former can be viewed as a "smoother version" of the latter since for any smooth even periodic function all its odd-order deviatives (not only the first one) coinside at $x = 0$ and $x = \pi$.

Denoting for any real s by Z_s^e a subspace of Z_s, formed by even functions, we observe that the equation (SG)+(EP) can be written in the Hamiltonian form (3.2) in the symplectic scale $(\{Z_s^e\}, \langle \bar{J} d\xi, d\xi \rangle)$, where $Z_s^e = \{\xi(x) \in Z_s \mid \xi$ satisfies (EP)$\}$. We note that for $s = 2$ the space Z_2^e is formed by the vector-functions from $H^2(0, \pi) \times H^2(0, \pi)$ which satisfy (N) (the functions are assumed to be extended to the segment $[0, 2\pi]$ in the even way). That is, for solutions of the SG equation in the Sobolev space H^2, the boundary conditions (OP) and (N) are equivalent.

Odd periodic boundary conditions. Similarly, the SG-equation under the odd periodic boundary conditions

$$u(t, x) \equiv u(t, x + 2\pi) \equiv -u(t, -x) \tag{OP}$$

can be written in a Hamiltonian form in the symplectic scale $(\{Z_s^o\}, \beta_2 = \langle \bar{J} d\xi, d\xi \rangle)$, where $Z_s^o = \{\xi(x) \in Z_s \mid \xi$ satisfies (OP)$\}$. These boundary conditions imply the Dirichlet ones: $(u_0, u_1')(0) = (u_0, u_1')(\pi) = (0, 0)$.

3.2 Lax-integrable equations

Let us consider a quasilinear Hamiltonian PDE and write it in the Hamiltonian form (3.1). This equation is called *Lax-integrable* if there exist linear operators $\mathcal{L}_u, \mathcal{A}_u$ which depend on $u \in Z_\infty$ and define linear morphisms of finite orders of some additional real or complex Hilbert scale $\{\mathfrak{Z}_s\}$, such that $u(t)$ is a smooth solution of (3.1) if and only if

$$\frac{d}{dt} \mathcal{L}_{u(t)} = [\mathcal{A}_{u(t)}, \mathcal{L}_{u(t)}]. \tag{3.4}$$

The operators $\mathcal{L}_u$ and $\mathcal{A}_u$ are said to form an $\mathcal{L}$-$\mathcal{A}$ pair of the equation (3.1)=(3.4).

We specify dependence of the $\mathcal{L}, \mathcal{A}$-operators on u and assume existence of integers s' and d' such that for all $s \geq s'$ the maps $u \mapsto \mathcal{L}_u$ and $u \mapsto \mathcal{A}_u$ are analytic as maps from Z_s to the space of bounded linear operators $\mathfrak{Z}_d \to \mathfrak{Z}_{d-d'}$, provided that $d \leq s$. Due to this assumption, the l.h.s of (3.4) is well defined for any C^1-smooth curve $u(t) \in Z_s$, $s \geq s'$.

Both KdV and Sine-Gordon equations are Lax-integrable, see below Section 4.

We abbreviate $\mathcal{L}_t = \mathcal{L}_{u(t)}$ and $\mathcal{A}_t = \mathcal{A}_{u(t)}$, where $u(t)$ is a smooth solution for (3.4). A crucial property of the $\mathcal{L}, \mathcal{A}$-operators is that spectrum of the operator $\mathcal{L}_t$ is time-independent and its eigen-vectors are preserved by the flow, defined by the operators $\mathcal{A}_t$:

Lemma 4. *Let $\chi_0 \in 3_\infty$ be a smooth eigenvector of $\mathcal{L}_0$, $\mathcal{L}_0\chi_0 = \lambda\chi_0$. Let us also assume that the initial-value problem*

$$\dot\chi = \mathcal{A}_t\chi, \quad \chi(0) = \chi_0, \tag{3.5}$$

has a unique smooth solution $\chi(t) \in 3_\infty$. Then $\mathcal{L}_t\chi(t) = \lambda\chi(t)$ for any t.

Proof. Let us denote $\mathcal{L}_t\chi(t) = \xi(t)$, $\lambda\chi(t) = \eta(t)$ and calculate derivatives of these functions. We have:

$$\frac{d}{dt}\xi = \frac{d}{dt}\mathcal{L}\chi = [\mathcal{A}, \mathcal{L}]\chi + \mathcal{L}\mathcal{A}\chi = \mathcal{A}\mathcal{L}\chi = \mathcal{A}\xi$$

and $\frac{d}{dt}\eta = \frac{d}{dt}\lambda\chi = \lambda\mathcal{A}\chi = \mathcal{A}\eta$. Thus, both $\xi(t)$ and $\eta(t)$ solve the problem (3.5) with χ_0 replaces by $\lambda\chi_0$ and coincide for all t by the uniqueness assumption. □

In many important examples of Lax-integrable equations, $\{Z_s\}$ is the Sobolev scale of L-periodic in x (vector-) functions and $\mathcal{L}, \mathcal{A}$ are u-dependent differential operators, acting on complex vector-functions. In this case it is natural to take for the scale $\{3_s\}$ the Sobolev scale of L-periodic complex vector-functions. So L-periodic spectrum of the operator $\mathcal{L}_u$ is an integral of motion for the equation (2.8) if the linear equation (2.11) defines a flow in the space of smooth L-periodic vector-functions. The set of integrals which we obtain in this way usually is incomplete. To get missing integrals we note that an L-periodic in x solution $u(t, x)$ can be also treated as an Lm-periodic solution for any $m \in \mathbb{N}$. Accordingly, we can consider the $\mathcal{L}, \mathcal{A}$-operators under mL-periodic boundary conditions and take for $\{3_s\}$ the Sobolev scale of mL-periodic vector-functions. Due to the lemma, the mL-periodic spectrum of $\mathcal{L}$ is an integral of motion if the equation (2.11) defines a flow in the corresponding space 3_∞. This set of integrals contains the initial one.

3.3   Integrable subsystems

Now we suppose that equation (3.1) possesses an invariant submanifold $\mathcal{T}^{2n} \subset O_d \cap Z_\infty$, such that restriction of the equation to $\mathcal{T}^{2n}$ is integrable. For some important examples the manifold $\mathcal{T}^{2n}$ may have singularities and the restricted symplectic form $\alpha_2 \mid_{\mathcal{T}^{2n}}$ may degenerate. Since our objects are analytic, these degenerations can only happen on singular subsets of positive codimension and do not affect the final KAM-results which neglect subsets of small measure: at some point we shall cut out the singular subsets with their small neighbourhoods. But our preliminary arguments are *global*. To carry them out we have to develop global notations.

We assume that $\mathcal{T}^{2n} = \Phi_0(R \times \mathbb{T}^n)$ where $\mathbb{T}^n = \{\jmath\}$ is the standard n-torus and R is a connected n-dimensional analytic set which is the real part of a connected complex analytic subset R^c of a domain $\Pi^c \subset \mathbb{C}^N$. [6] By R_s^c

we denote any proper analytic subset of R^c which contains its singular part and denote by R_s the real part of R_s^c, i.e., $R_s = R_s^c \cap R$.

We assume that the map Φ_0 is analytic and the form $\alpha_2 \mid_{\mathcal{T}^{2n}}$ does not degenerate identically:

(i) The map $\Phi_0 : R \times \mathbb{T}^n \to Z_l$ is analytic for each l. That is, for some $\delta = \delta_l > 0$ it extends to an analytic map $\Pi^c \times \{|\mathrm{Im}\,\mathfrak{z}| < \delta\} \to Z_l^c$.

(ii) R^c contains a proper analytic subset R_{s1}^c such that the analytic 2-form $\Phi_0^* \alpha_2$ is non-degenerate in $(R \setminus (R_s \cup R_{s1})) \times \mathbb{T}^n$, where $R_{s1} = R_{s1}^c \cap R$.

For brevity we re-denote $R_s := R_s \cup R_{s1}$ and similar re-denote R_s^c. We set $R_0^c = R^c \setminus R_s^c$ and $R_0 = R \setminus R_s$. Since R_s and R_s^c comprise singularities of the analytic sets R and R^c as well as of the map Φ_0, then the sets R_0 and R_0^c are smooth analytic manifolds and the map $\Phi_0 : R_0 \times \mathbb{T}^n \to Z_l$, $\Phi_0(R_0 \times \mathbb{T}^n) =: \mathcal{T}_0^{2n}$, is an immersion. We assume that

(iii) The set $\mathcal{T}_0^{2n}$ is a smooth analytic submanifold of each space Z_l, invariant for the equation (3.1), as well as the tori $T^n(r) = \Phi_0(\{r\} \times \mathbb{T}^n)$, $r \in R_0$. The restricted to $T^n(r)$ equation takes the form $\dot{\mathfrak{z}} = \omega(r)$, where ω extends to an analytic map $\Pi^c \to \mathbb{C}^n$.

Due to (ii) and (iii), the manifold $\mathcal{T}_0^{2n}$ is filled with smooth time-quasiperiodic solutions of the equation (3.1).

The *frequency map* $r \mapsto \omega(r)$ is assumed to be non-degenerate:

(iv) for almost all $r \in R_0$, the tangent map $\omega_*(r) : T_r R_0 \to \mathbb{R}^n$ is an isomorphism.

By Theorem 2 the equation restricted to $\mathcal{T}_0^{2n}$ is Hamiltonian. Condition iv) implies that this equation is integrable (see [KK] or [BoK2], p. 106). So by the Liouville–Arnold theorem, R_0 can be covered by a countable system of open domains R_{0j}, $R_0 = R_{01} \cup R_{02} \cup \cdots$, such that the equation (3.1) restricted to each manifold $\mathcal{T}_j^{2n} = \Phi_0(R_{0j} \times \mathbb{T}^n)$ admits action-angle variables (p, q) with actions $p \in P_j \Subset \mathbb{R}^n$ and angles $q \in \mathbb{T}^n$. I.e., $\omega_2 = dp \wedge dq$ and the equation restricted to $\mathcal{T}_j^{2n}$ takes the form

$$\dot{p} = 0, \quad \dot{q} = \nabla h(p), \quad h = \mathcal{H} \circ \Phi_0,$$

where $h = h_j$.

Lemma 5. *In any domain R_{0j} we have $\nabla h(p(r)) = \omega(r)$ and $q = \mathfrak{z} + q^0(r)$.*

For many important examples the set of integrals of motion of a Lax-integrable equation, formed by 2π-periodic and 2π-antiperiodic spectra (see Lemma 4 and its discussion), is "complete". Fixing all but some n of them one can construct invariant manifolds $\mathcal{T}^{2n}$ as above. Below we show how to carry out this construction for KdV and SG equations.

---

[6] That is, R^c is formed by zeroes of an analytic map $\Pi^c \to \mathbb{C}^{N-n}$ such that at some points of Π^c its linearisation has full rank. For elementary facts concerning analytic sets, real and complex, see [Mil] and [GR], Sections II, III.

4 Finite-gap manifolds and theta-formulas

We start with famous finite-gap solutions for the KdV equation under zero-meanvalue periodic boundary conditions:

$$\dot{u} = \frac{1}{4}\frac{\partial}{\partial x}(u_{xx} + 3u^2), \quad u(t,x) \equiv u(t, x + 2\pi), \quad \int_0^{2\pi} u\,dx \equiv 0. \quad \text{(KdV)}$$

The finite-gap solutions fill invariant submanifolds $\mathcal{T}^{2n} \subset H_0^s(S^1)$ with integrable dynamics on them, as in Section 2. To study the manifolds $\mathcal{T}^{2n}$ we shall use the Its - Matveev formula which represents the finite-gap solutions in terms of theta-functions.

4.1 Finite-gap manifolds for the KdV equation

The $\mathcal{L}, \mathcal{A}$-operators for the KdV equation are $\mathcal{L}_u = -\frac{\partial^2}{\partial x^2} - u$ and $\mathcal{A}_u = \frac{\partial^3}{\partial x^3} + \frac{3}{2}u\frac{\partial}{\partial x} + \frac{3}{4}u_x$. Indeed, calculating the commutator $[\mathcal{A}, \mathcal{L}]v$ one sees that most of the terms cancel and there is nothing left except $(\frac{1}{4}u_{xxx} + \frac{3}{2}uu_x)v$. Thus, $[\mathcal{A}, \mathcal{L}]$ is an operator of multiplication by the r.h.s. of KdV and the equation can be written in the form (3.4). For the scale $\{\mathfrak{Z}_s\}$ we take one of the following scales of complex functions: or the scale of 2π-periodic functions, or the scale of 2π-antiperiodic functions, or the scale of 4π-periodic ones.

It is well-known [Ma, MT] that the spectrum of the Sturm–Liouville operator $\mathcal{L}_u$ acting on twice differentiable functions of period 4π is a sequence of simple or double eigenvalues $\{\lambda_j \mid j \geq 0\}$, tending to infinity:

$$\lambda_0 < \lambda_1 \leq \lambda_2 < \lambda_3 \leq \lambda_4 < \cdots \nearrow \infty.$$

Corresponding eigenfunctions are smooth if the potential $u(x)$ is. The spectrum $\{\lambda_j\}$ can be also described without doubling the period: it equals the union of the periodic and antiperiodic spectra of the operator $\mathcal{L}_u$, considered on the segment $[0, 2\pi]$. Below we denote $\lambda = \{\lambda_0, \lambda_1, \dots\}$ and refer to the sequence $\lambda = \lambda(u)$ as to the periodic/antiperiodic spectrum of the operator $\mathcal{L}_u$.

The segment $\Delta_j = [\lambda_{2j-1}, \lambda_{2j}]$, $j = 1, 2, \dots$, is called the jth spectral gap. The gap Δ_j is *open* if $\lambda_{2j} > \lambda_{2j-1}$ and is *closed* if $\lambda_{2j} = \lambda_{2j-1}$.

Example. For $u = 0$ we have $\lambda_{2k} = k^2/4$, $k \geq 0$, and $\lambda_{2l-1} = l^2/4, l \geq 1$. Corresponding eigen-functions are $(2\pi)^{-1/2}\cos kx/2$ and $(2\pi)^{-1/2}\sin lx/2$. All gaps Δ_j are now closed. $\square$

If $u(t, x)$ is a smooth x-periodic function, then the linear equation $\dot{v} = \mathcal{A}_{u(t,x)}v$, $v(0, x) = v_0(x)$, has a unique smooth periodic solution $v(t, x)$ for any given smooth periodic initial data $v_0(x)$ (this follows from an abstract theorem in [Paz]). Hence, Lemma 4 with $\{\mathfrak{Z}_s = H^s(\mathbb{R}/4\pi\mathbb{Z})\}$ imply that the sequence λ is an integral of motion:

$$\lambda(u(t, \cdot)) \text{ is time-independent if } u(t, x) \text{ is a solution of the KdV.} \quad (4.1)$$

Let us fix any integer n-vector $\Upsilon = (V_1, \ldots, V_n) \in \mathbb{N}^n$, where $V_1 < \cdots < V_n$, and consider a set $\mathcal{T}_\Upsilon^{2n}$,

$$\mathcal{T}_\Upsilon^{2n} = \{u(x) \mid \text{the gap } \Delta_j(u) \text{ is open iff } j \in \{V_1, \ldots, V_n\}\}.$$

This set equals to the union of isospectral subsets $T^n(r) = T_\Upsilon^n(r)$ with prescribed lengths of the open gaps: $\mathcal{T}_\Upsilon^{2n} = \bigcup_{r \in \mathbb{R}_+^n} T_\Upsilon^n(r)$, where $T_\Upsilon^n(r) = \{u(x) \in \mathcal{T}_\Upsilon^{2n} \mid |\Delta_{V_j}| = r_j \; \forall j\}$. By (4.1) each set $T_\Upsilon^n(r)$ is invariant for the KdV-flow.

Remarkably, the whole spectrum λ of an n-gap potential is defined by the n-vector r and analytically depends on it [GT]. Each set $T_\Upsilon^n(r)$ is not empty and is an analytic n-torus in any space $H_0^s = H_0^s(S^1)$. The tori $T_\Upsilon^n(r)$ are analytically glued together, so $\mathcal{T}_\Upsilon^{2n}$ is an analytic submanifold of each space H_0^s. – These are well-known results from the inverse spectral theory of the Sturm-Liouville operator $\mathcal{L}_u$, see [Ma] and [Mo, GT, MT]. So the inverse spectral theory provides us with KdV-invariant $2n$-manifolds foliated to invariant n-tori.

4.2   The Its–Matveev theta-formulas

To check that the n-gap manifolds $\mathcal{T}_\Upsilon^{2n}$ of the KdV equation possesses the properties (i)–(iv) from Section 2, we have to present an analytic map Φ_0 as in Section 2 and to check its properties. We shall write the map Φ_0 in terms of theta-functions, following [D, BB].

Let us take any n-gap potential $u(x) \in T_\Upsilon^n(r)$ and denote by $E_1(r) < E_2(r) < \cdots < E_{2n+1}$ end points of the open gaps plus λ_0 (so $E_1 = \lambda_0$ and $\Delta_{V_1} = [E_2, E_3], \ldots, \Delta_{V_n} = [E_{2n}, E_{2n+1}]$). The Riemann surface $\Gamma = \Gamma(r)$ of genus n,

$$\Gamma = \{P = (\lambda, \mu) \mid \mu^2 = R(\lambda; r) := \prod_{j=1}^{2n+1} (\lambda - E_j(r))\},$$

has branching points at $E_1, \ldots, E_{2n+1}$ and ∞. Let $a_1, \ldots, a_n$ be the ovals in Γ lying above the open gaps $\Delta_{V_1}, \ldots, \Delta_{V_n}$ (i.e., $a_j = \pi^{-1} \Delta_{V_j}$). After Γ is cut along these ovals, it falls to two sheets Γ_+, Γ_-, chosen in such a way that μ is positive on the upper edge of the cut $[E_{2n+1}, E_\infty]$ in Γ_+.

We supplement the ovals a_j by n b-circles $b_1, \ldots, b_n$ in such a way that the circles have the canonical intersection matrix: $a_i \circ a_j = b_i \circ b_j = 0$ and $a_i \circ b_j = \delta_{ij}$. Next we take a basis $d\omega_1, \ldots, d\omega_n$ of holomorphic differentials on Γ, normalised by the conditions $\langle d\omega_j, a_k \rangle := \oint_{a_k} d\omega_j = 2\pi i \delta_{jk}$. These differentials exist and are uniquely defined by the normalisation.

The Riemann matrix $B = B(r) = (B_{jk})$ of the curve Γ is defined as the matrix of b-periods of the differentials $d\omega_j$:

$$B_{jk} = \langle d\omega_j, b_k \rangle.$$

Under our choice of the a, b-cycles, the matrix B is real (see [BB, KK]). Its symmetric part is negatively defined due to general properties of the Riemann matrices.

Now we define the theta-function θ of the curve $\Gamma = \Gamma(r)$:

$$\theta = \theta(z; r) = \sum_{s \in \mathbb{Z}^n} \exp\left(\frac{1}{2}(B(r)s, s) + (z, s)\right),$$

where $z \in \mathbb{C}^n$ (the sum converges due to the properties of the Riemann matrix B). Clearly the function is 2π-periodic in imaginary directions: $\theta(z+2\pi i e_k) = \theta(z)$, where e_k is the k-th basis vector of $\mathbb{C}^n$. The differentials $d\omega_j$ analytically depend on the parameter $r \in \mathbb{R}^n_+$ as well as the matrix $B(r)$. Therefore the function $\theta(z; r)$ is analytic in $r \in \mathbb{R}^n_+$.

Since the matrix B is real, then θ is real and even: $\overline{\theta(z)} = \theta(\bar{z})$ and $\theta(z) = \theta(-z)$. In particular, this function is real both in real and pure imaginary directions.

Next, on the surface $\Gamma(r)$ we consider Abelian differentials of the second kind $d\Omega_1, d\Omega_3$ with vanishing a-periods and with the only poles at infinity of the form

$$
\begin{aligned}
d\Omega_1 &= dk + (c + O(k^{-2})) \, dk^{-1}, \quad k = i\sqrt{\lambda} \to \infty, \\
d\Omega_3 &= dk^3 + O(1) \, dk^{-1},
\end{aligned}
\tag{4.2}
$$

where c is an unknown constant. The normalisation (4.2) defines the differentials uniquely, see [S, NMP, BB]. Besides,

$$
d\Omega_1 = \frac{i}{2} \frac{\lambda^n + \cdots}{\mu} \, d\lambda, \quad d\Omega_3 = -\frac{3}{2} i \frac{\lambda^{n+1} + \cdots}{\mu} \, d\lambda,
\tag{4.2'}
$$

where the dots stand for real polynomials of degree $n - 1$.

Let us define complex n-vectors $i\mathbf{V}(r)$ and $i\mathbf{W}(r)$ as the vectors of b-periods of these differentials:

$$iV_j = \langle d\Omega_1, b_j \rangle, \quad iW_j = \langle d\Omega_3, b_j \rangle.$$

The vector $\mathbf{V}$ is called the *wave-number vector* and $\mathbf{W}$ – the *frequency vector*.

The vector $\mathbf{W}$ is real and the vector V is integer. Moreover, V equals to the vector, formed by the numbers of the open gaps, see [BB, KK] (the latter vector was earlier denoted also as V).

One of top achievements of the finite-gap theory is the Its–Matveev formula, which represents any n-gap potential $u(x) \in T^n(r)$ in the form

$$u(x) = u(x; r, \mathfrak{z}) = 2\frac{\partial^2}{\partial x^2} \ln \theta(i\mathbf{V}x + i\mathfrak{z}; r) + 2c.
\tag{4.3}$$

Here the constant c is the same as in (4.2). Since the mean-value of the r.h.s. in (4.3) equals $2c$, then we must have $c = 0$. The vector $i\mathfrak{z}$ in (4.3) is $i_3 = -A(\mathcal{D}) - \mathbf{K}$, where $\mathbf{K}$ is the vector of Riemann constants (see [D, BB]) and $A(\mathcal{D})$ is the Abel transformation of a positive divisor $\mathcal{D} = \mathcal{D}(\sqcap)$, $\mathcal{D} = \mathcal{D}_\infty \ldots \mathcal{D}_\backslash$, $D_j \in a_j$. I.e., $A(\mathcal{D})$ is a complex n-vector such that its jth component $A(\mathcal{D})_|$ equals

$$A(\mathcal{D})_j = \sum_{r=1}^{n} \int_{\infty}^{D_r} d\omega_j,$$

where $\{d\omega_j\}$ are the holomorphic differentials on Γ as above. The divisor $\mathcal{D}$ is a divisor of Dirichlet eigenvalues, i.e. $D_j = (\lambda_j, \mu_j)$, where λ_j is an eigenvalue of the operator $\mathcal{L}_u$ subject to Dirichlet boundary conditions $\varphi(0) = \varphi(2\pi) = 0$ (each gap Δ_j contains exactly one point from the Dirichlet spectrum, see [Ma, MT]).[7] In particular, every point D_j analytically depends on u.

The phase vector $_3$ turns out to be real, so $\theta(i\mathbf{V}x + i_3)$ is a real valued function of x. The theta-function is nonzero at any imaginary point $i\xi \in i\mathbb{R}^n$ (see in [BB]). Since this function is periodic, then $|\theta(i\xi)| \geq C(r) > 0$ for any real vector ξ. Hence, the r.h.s. of (4.3) is analytic in $_3 \in \mathbb{T}^n$.

Due to the periodicity, we can treat $_3$ as a point in the torus $\mathbb{T}^n$. Thus we get an analytic map:

$$T^n(r) \to \mathbb{T}^n, \quad u(\cdot) \mapsto {}_3.$$

This map has the analytic inverse given by the formula (4.3). The coordinate $_3$ on a finite-gap torus $T^n(r)$ is called the *theta-angles*.

Time-evolution $u(t, x)$ of the n-gap potential $u(x) \in T^n(r)$ as in (4.3) along the KdV flow is given by the following formula, also due to Its–Matveev:

$$u(t, x; r, {}_3) = 2\frac{\partial^2}{\partial x^2} \ln \theta(i(\mathbf{V}x + \mathbf{W}t + {}_3); r) \qquad (4.4)$$

(we use that $c = 0$).

Denoting by G the function $G({}_3, r) = 2\frac{\partial^2}{\partial x^2} \ln \theta(i\mathbf{V}x + i_3; r)\,|_{x=0}$, we write u as $u(x; r, {}_3) = G({}_3 + \mathbf{V}x, r)$. So the n-gap torus $T^n(r)$ is represented in the form

$$T^n(r) = \Phi_0(r, \mathbb{T}^n) \subset H_0^d,$$

where for any $_3 \in \mathbb{T}^n$, $\Phi_0(r, {}_3)$ is the analytic function $\Phi_0(r, {}_3)(x) = G(\mathbf{V}x + {}_3, r)$. The formula (4.4) can be written as

$$u(t, x; r, {}_3) = \Phi_0(r, {}_3 + \mathbf{W}(r)t)(x).$$

This shows that in the $(r, {}_3)$-variables the KdV-flow on $\mathcal{T}^{2n}$ takes the form $\dot{_3} = \mathbf{W}(r)$. I.e., the theta-angles $_3$ integrate the KdV-equation on any torus $T^n(r)$.

---

[7] This divisor can be also described as a divisor of poles of the Baker–Akhiezer eigenfunction $\varphi(x; P)$ of the operator $\mathcal{L}_u$, $\mathcal{L}_u\varphi = \pi(P)\varphi$, normalised at infinity as $\varphi \sim e^{i\sqrt{\lambda}x}$; see [D, BB].

Let R be a sub-cube of the octant $\mathbb{R}^n_+$ of the form $R = \{r \in \mathbb{R}^n_+ \mid 0 < r_j < K\}$ with some $K > 0$, and

$$\mathcal{T}^{2n} = \Phi_0(R \times \mathbb{T}^n) \subset \mathcal{T}^{2n}_{\mathbf{V}} \tag{4.5}$$

for any wave-number vector $\mathbf{V}$. The set $\mathcal{T}^{2n} \subset H^d_0$, $d \geq 1$, is an invariant manifold of the KdV equation. It meets the assumptions i)–iii) from Section 2 since: The map Φ_0 is an analytic embedding and $\mathcal{T}^{2n}$ is an analytic submanifold of H^d_0. The form $\Phi^*_0 \alpha_2$ is analytic and is non-degenerate for small r since it is nondegenerate at $r = 0$ by Theorem 3.1, so the set of its degeneracy is a proper analytic subset of the cube R (in fact, it is empty).

The non-degeneracy assumption iv) also holds, as states the following Non Degeneracy Lemma:

Lemma 6. *The determinant* $\det\{\partial W_j / \partial r_k\}$ *is nonzero almost everywhere.*

This result can be proven directly [BiK2, KK], or can be obtained as an immediate consequence of another lemma (proven in [BoK1, KK]):

Lemma 7. *For any finite-gap manifold* $\mathcal{T}^{2n}_{\mathbf{V}}$ *the corresponding frequency vector* $\mathbf{W}$ *as a function of* r *analytically extends to the origin and has there the following asymptotic:* $W_j(r) = -\frac{1}{4}V^3_j + \frac{3}{8\pi V_j}r^2_j + O(|r|^4)$, $j = 1, \ldots, n$.

4.3   Higher equations from the KdV hierarchy

Let us take any n-gap manifold $\mathcal{T}^{2n}_{\mathbf{V}}$. The manifold itself and each torus $T^n(r) \subset \mathcal{T}^{2n}_{\mathbf{V}}$ are invariant for all Hamiltonian equations with the hamiltonians $\mathcal{H}_0$, $\mathcal{H}_1, \ldots$ from the KdV-hierarchy (see Example 3.2). The flow of any l-th KdV equation on $\mathcal{T}^{2n}_{\mathbf{V}}$ is very similar to the KdV-flow: it is given by the theta-formula (4.4) where the frequency-vector $\mathbf{W}$ should be replaced by an n-vector $\mathbf{W}^{(l)}$, formed by b-periods of some Abelian differential $d\Omega_{2l+1}$. All results till Lemma 6, stated above for the KdV equation, have obvious reformulations for the higher KdV's. An analogy for Lemma 7 is given by the following statement: The vector $\mathbf{W}^{(l)}$ is analytic in $r^2_1, \ldots, r^2_n$ and

$$W^{(l)}_j(r) = W^{(l)}_{j0} + W^{(l)}_{j1}r^2 + O(|r|^4) \tag{4.6}$$

for any $j = 1, \ldots, n$, with some non-zero constants $W^{(l)}_{j1}$. Any manifold $\mathcal{T}^{2n}_{\mathbf{V}}$ treated as an invariant manifold of an lth KdV equation satisfies assumptions i)-iv) for the same reason as for $l = 1$ (i.e., as in the KdV-case).

4.4   Sine-Gordon equation under Dirichlet boundary conditions

Similar to the KdV-case, the equation (SG)+(EP) (see Example 3.3) has time-quasiperiodic finite-gap solutions with n basic freqencies which can be

written in terms of theta-functions. To symplify presentation we restrict ourselves to finite-gap solutions with even number of gaps, so

$$g = 2n$$

everywhere below. These solutions can be written as $\xi = (u, v)(t, x; D, r)$,
where $v = -\dot{u}$ and

$$u(t, x) = 2i \log \frac{\theta(i(Vx + Wt + D + \Delta))}{\theta(i(Vx + Wt + D))}. \tag{4.7}$$

Here θ is a theta-function of the Riemann surface Γ,

$$\Gamma = \Gamma(r) = \{(\lambda, \mu) \mid \mu^2 = \lambda \prod_{j=1}^{n} (\lambda - E_j)(\lambda - \bar{E}_j)(\lambda - E_j^{-1})(\lambda - \bar{E}_j^{-1})\},$$

Δ is the vector $(\pi, \ldots, \pi)$, the wave-number vector V and the frequency
vector W are constructed in a way, similar to the KdV-case and

$$\mathcal{D} = (D_1, \ldots, D_n, D_n, D_{n-1}, \ldots, D_1), \quad D = (D_1, \ldots, D_n) \in \mathbb{T}^n.$$

The vector $r = (E_1, \ldots, E_n)$ is a point from some connected n-dimensional
real algebraic set R, which is a bounded part of the set $\mathbb{C}_+^n$, $\mathbb{C}_+ = \{x + iy \mid y > 0\}$. The algebraic set R is "smooth near the real subspace" in the sence
that for some $\delta_0 > 0$ the set $R_+ = \{(E_1, \ldots, E_n) \in R \mid 0 < E_j < \delta_0 \, \forall j\}$ is a
smooth n-dimensional real analytic manifold.

Since the set of branching points of the surface Γ is inversion-invariant,
then the vector W (the vector V) turned out to be symmetric (antisymmet-
ric) with respect to the involution T, $T(U_1, \ldots, U_{2n}) = (U_{2n}, \ldots, U_1)$:

$$TW = W, \qquad TV = -V.$$

Hence, W and V are determined by the vectors $\tilde{W}$ and $\tilde{V}$, formed by the
last n components:

$$\tilde{W} = (W_{n+1}, \ldots, W_{2n}), \qquad \tilde{V} = (V_{n+1}, \ldots, V_{2n}).$$

These vectors also will be called the frequency and wave-number vectors.

The wave-number vector $\tilde{V}(r)$ is an r-independent integer vector (this
guarantees that u is 2π-periodic in x). An important property of the family
of solutions we discuss is that the closure $\bar{R}$ contains an unique real point $r^0 = (E_1^0, \ldots, E_n^0) \in \mathbb{R}^{2n}$ and $u \to 0$ when $r \to r^0$ (this convergence corresponds
to a degeneration of the Riemann surface Γ when the gaps $[E_j, \bar{E}_j]$ shrink to
points). Moreover,

$$u(0, x; D, r) = \sum_{j=1}^{n} \varepsilon_j \Big(\cos(V_{n+j} x + D_j) + \cos(-V_{n+j} x + D_j) \Big) + o(|\varepsilon|^2),$$

$$\tag{4.8}$$

where $\varepsilon \to 0$ as $r \to r_0$. For the sake of simplicity we restict ourselves to finite-gap solutions such that first n gaps are open in the sence that

$$(V_{n+1}, \ldots, V_{2n}) = (1, \ldots, n). \tag{4.9}$$

The frequency vector $\tilde{W}$ is analytic on the algebraic set R and satisfies the following analogies of Lemmas 6 and 7:

a) the tangent map $\tilde{W}(r)_*$ is nondegenerate for a.a. $r \in R$;

b) the set R_+ admits analytic coordinates $\mu_1, \ldots, \mu_n$ such that $\text{Im}\, E_j \to 0$ as $\mu_j \to 0$, the map $\tilde{W}$ is analytic in μ and

$$W_{n+j}(0) = \sqrt{1 + j^2} =: j^*,$$
$$\frac{\partial W_{n+j}}{\partial \mu_k}\Big|_{\mu=0} = \begin{cases} -16/j^*, & j \le n, j \ne k, \\ -12/j^*, & j = k. \end{cases} \tag{4.10}$$

The set of time-quasiperiodic solutions (4.7) is incomplete in the sence that the infinitesimal solutions (4.8) do not contain the "$\cos 0x$" term $\varepsilon_0 \cdot 1$. The reminding solutions are odd-gap, they can be treated similar.

For all these results see [BB], [BiK1, BoK2] and [KK].

A solution (4.7) is winding in a torus

$$T^n(r) = \{\xi(0, \cdot; D, r) \mid D \in \mathbb{T}^n\}.$$

Altogether the tori form a set $\mathcal{T}^{2n} = \bigcup_{r \in R} T^n(r)$ which is an image of $\mathbb{T}^n \times R$ under a map Φ_0, defined in terms of the r.h.s. of (4.7) as in the KdV-case. The set $\mathcal{T}^{2n}$ and the map Φ_0 satisfy the assumptions (i)–(iii) for the same trivial reasons as in the KdV-case, while assumption iv) follows from a).

5   Linearised equations and their Floquet solutions

5.1   The linearised equation

Below (Z, α_2) stands for a symplectic space $(Z = Z_d, \alpha_2 = \bar{J}\, dz \wedge dz)$ with some fixed d. We continue to study a quasilinear Hamiltonian equation

$$\dot{u} = J\nabla\mathcal{H}(u) = J(Au + \nabla H(u)) =: V_\mathcal{H}(u), \tag{5.1}$$

where $\text{ord}\, A = d_A$, $\text{ord}\, \nabla H = d_H < d_A$, $\text{ord}\, J = d_J$ and $d \ge d_A/2$. The equation is assumed to possess a $2n$-dimensional invariant manifold $\mathcal{T}^{2n} \subset Z$ as in Section 3, satisfying the assumptions i)-iv). By iii), the regular part $\mathcal{T}_0^{2n}$ of $\mathcal{T}^{2n}$ is filled by smooth solutions $u_0(t)$ of (5.1) of the form $u_0(t) = u_0(t; r_0, \mathfrak{z}_0) = \Phi_0(w_0(t))$, where $w_0(t) = (r_0, \mathfrak{z}_0 + tw(r_0)) \in R_0 \times \mathbb{T}^n$, $t \in \mathbb{R}$. We linearise (5.1) about a solution u_0 as above to get the nonautonomous linear equation

$$\dot{v} = J(Av + \nabla H(u_0(t))_* v) =: JA_t(t)v, \tag{5.2}$$

which is our concern in this section. We recall that linear flow-maps of equation (5.2) (if they exist) are denoted as $S^t_{\tau*}(u_0(\tau))$ (see Definition 1), and supplement the assumptions i)–iv) assuming that

(v) for any solution u_0 of (5.1) in $\mathcal{T}_0^{2n}$ the flow-maps $S^t_{\tau*}(u_0(\tau))$, $-\infty < \tau, t < \infty$, are well defined in the space $Z = Z_d$.

Since the equation (5.1) is Hamiltonian, then the flow-maps $S^t_{\tau*}(u)$ $(u \in \mathcal{T}_0^{2n})$ are symplectomorphisms of the symplectic space (Z, α_2) (see [KK]).

To study equation (5.1) near $\mathcal{T}^{2n}$ we shall impose an integrability assumption on the linearised equation (5.2). Roughly speaking, this assumption means that the equation (5.2) has a complete system of time-quasiperiodic Floquet solutions.

Since the time-flow of (5.2) is formed by linear symplectomorphisms which preserve tangent spaces to $\mathcal{T}_0^{2n}$, then the flow also defines symplectomorphisms of skew-orthogonal complements $T_{u_0}^\perp \mathcal{T}_0^{2n}$ to the spaces $T_{u_0} \mathcal{T}_0^{2n}$ in $T_{u_0} Z \sim Z$.[8]

5.2 Floquet solutions

We call a solution $v(t)$ of the equation (5.2) a Floquet solution if there exists a section Ψ of the complexified tangent bundle to Z, restricted to $\mathcal{T}_0^{2n}$ (i.e., $R_0 \times \mathbb{T}^n \ni (r, \mathfrak{z}) \mapsto \Psi(r, \mathfrak{z}) \in T_{\Phi_0(r,\mathfrak{z})} \mathcal{T}_0^{2n}$), and a complex function $\nu(r)$ such that the solution v has the form

$$v(t) = v(t; r_0, \mathfrak{z}_0) = e^{i\nu(r_0)t} \Psi(w_0(t)), \quad w_0 = (r_0, \mathfrak{z}_0 + t\omega(r_0)). \qquad (5.3)$$

It is assumed that $v(t)$ solves (5.2) for any choice of $r_0 \in R_0$ and $\mathfrak{z}_0 \in \mathbb{T}^n$. We call $\nu(r)$ the (Floquet) exponent of a Floquet solution v.

A Floquet solution $v(t)$ is called a *skew-orthogonal Floquet solution* if Ψ in (5.3) is a section of the complexified skew-normal bundle $T^{\perp c}\mathcal{T}^{2n}$ (its fibres are complexifications of the spaces $T_{u_0}^\perp \mathcal{T}_0^{2n}$).

Let us assume that equation (5.2) has an infinite family of Floquet solutions $v = v_j(t)$. Clearly if v_j is a Floquet solution, then $\overline{v}_j$ is a solution with the exponent $-\overline{\nu}_j(r)$, corresponding to the section $\overline{\Psi}_j$. We add this solution to the family; if a solution with the exponent $-\overline{\nu}(r)$ already was there, we replace it by $\overline{v}_j$. Now the family is invariant with respect to the complex conjugation and the set of all exponents is invariant with respect to the involution $\nu \to -\overline{\nu}$. In addition we *suppose* that the set of exponents is invariant with respect to the complex conjugation $\nu \to \overline{\nu}$ (this assumption holds trivially if all the frequencies are real); hence the set is invariant with respect to the involution $\nu \to -\nu$.

It is convenient to enumerate the Floquet solutions by integers from the set $\mathbb{Z}_n = \{\pm(n+1), \pm(n+2), \dots\}$. We do it in such a way that $\nu_{-j}(p) \equiv -\nu_j(p)$

---

[8] $T_{u_0}^\perp \mathcal{T}_0^{2n}$ is formed by all vectors $\xi \in T_{u_0} Z$ such that $\alpha_2(\xi, \eta) = 0$ for each $\eta \in T_{u_0} \mathcal{T}_0^{2n}$.

and $\Psi_{-j} \equiv \overline{\Psi}_j$ if ν_j is real. So below we consider the following system of Floquet solutions :

$$v_j(t; r_0, \mathfrak{z}_0) = e^{i\nu_j(r_0)t} \Psi_j(r_0, \mathfrak{z}_0 + t\omega(r_0)), \quad j \in \mathbb{Z}_n; \quad \nu_{-j}(r) \equiv -\nu_j(r). \quad (5.4)$$

For each index k we denote by $\hat{k}$ an index such that $\nu_{\hat{k}} = \overline{\nu_k}$. Clearly $\hat{\hat{k}} = k$ for any k and $\hat{k} = k$ if ν_k is real. We note that the hat-map is r-independent in any connected sub-domain of R_0 where all the functions $\nu_j(r)$ are different.

A frequency ν_j can have an algebraic singularity at the set Λ_j formed by all r such that $\nu_j(r) = \nu_p(r)$ for some $p \neq j$. Situation becomes too intricate if there are infinitely many nontrivial sets Λ_j. To avoid this complexification we assume that

a) there is a nonempty sub-domain of R_0 where $\nu_j \neq \nu_k$ if $j \neq k$. Moreover, there exists j_1 (depending on $\mathcal{T}^{2n}$) such that $\nu_j(r) \neq \nu_k(r)$ for all r, all k and all $|j| \geq j_1$, $j \neq k$.

Since $\nu_{-j} = -\nu_j$, then by this assumption $\nu_j \neq 0$ if $j \geq j_1$.

The exponents ν_k with $|k| \geq j_1$ are assumed to be real analytic:

b) for any k such that $|k| \geq j_1$, ν_k is a real-valued analytic function on R (so $\nu_k \equiv -\nu_{-k}$ and $\Psi_{-k} = \overline{\Psi}_k$). The section Ψ_k extends to an analytic map $\Pi^c \times \{|\operatorname{Im}\mathfrak{z}| < \delta\} \to Z^c$ and ν_k extends to an analytic function on Π^c.

In particular, $\hat{k} = k$ if $|k| \geq j_1$.

For sophisticated integrable equation like the SG equation some exponents $\nu_k(r)$ with $|k| < j_1$ have non-trivial algebraic singularities. The assumptions we shall impose now on the exponents ν_k with $|k| < j_1$ (below we call such k small), are made ad hoc – they are met by Floquet solutions of Lax-integrable equations. These assumptions are empty for integrable equations with seladjoint $\mathcal{L}$-operators, e.g., for the KdV equation.

Let us denote $M = j_1 - n - 1$.

c) The functions ν_j with small j are continuous in R and have the form $\nu_j(r) = \tilde{\nu}(\lambda_j(r), r)$, where $\{\lambda_j(r)\}$ are branches of some $2M$-valued algebraic function, defined on Π^c, and $\tilde{\nu}$ is an analytic complex function.

By c) the functions ν_j with small j are algebraic (as well as the functions λ_j). They are analytic in Π^c outside a discriminant set D of the algebraical function $\tilde{\nu}$. We note that $D \cap R$ is a proper analytic subset of R since no two exponents ν_j, ν_k coincide identically in R by the assumption i).

The set $D \cap R^c$ contains algebraic singularities of the Floquet exponents and is contained in the set

$$\Lambda = \{r \in R^c \mid \nu_j(r) = \nu_k(r) \text{ for some } j \neq k, \text{ with } |j|, |k| < j_1\},$$

which is a proper analytic subset of R^c. It contains zeroes of the exponents ν_j since they are odd in j. We add Λ to the singular set R_s^c:

$$R_s^c := R_s^c \cup (\Lambda \cap R^c), \quad R_s := R_s \cup (\Lambda \cap R),$$

and modify the regular set R_0 accordingly.

Example. Eigenvalues $\{\lambda_j\}$ of a real matrix $B(a)$ which analytically depends on a real vector-parameter a are zeroes of the characteristic equation $\det(B(a) - \lambda E) = 0$ and are algebraic functions of a. A priori they have singularities at the sets $\Lambda_{jk} = \{\lambda_j = \lambda_k\}$. Some of these singularities can be removed by re-enumerating the eigenvalues before or behind the sets Λ_{jk}. In particular, if the matrix $B(a)$ is symmetric, then under proper enumeration the eigenvalues have no singularities at all (this is Rellich's theorem, see [Kat2,RS4]). However, if λ_j and λ_k are real "before" Λ_{jk} and have nontrivial imaginary parts "behind" Λ_{jk}, then a singularity at this set is unremovable.

Now we pass to smoothness of the sections Ψ_j with small j ($|j| < j_1$):

d) There exists a bounded analytic map $(r, \jmath, \lambda) \mapsto \widetilde{\Psi}(r, \jmath, \lambda) \in Z^c$, such that $\Psi_j(r, \jmath) = \widetilde{\Psi}(r, \jmath; \lambda_j(r))$ for $(r, \jmath) \in R_0 \times \mathbb{T}^n$ and all small j.

This assumption agrees with smoothness of eigenvectors in finite-dimensional spectral problems:

Example (continuation). Let us denote by $B^j = (B^j_{lm} \mid 1 \le l, m \le n)$ the matrix $B^j(a) = B - \lambda_j(a)E$, so $B\xi = \lambda_j\xi$ if $B^j\xi = 0$. Let us assume that $\operatorname{rk} B^j(a) = n - 1$ for $a \notin \Lambda_j = \bigcup_k \Lambda_{jk}$ and denote by $\xi_m(a)$ the algebraic complement to the element $B^j_{nm}(a)$ in the matrix B^j. Then the vector $\xi = (\xi_1, \ldots, \xi_n)$ is nonzero for $a \notin \Lambda_j$ and $\sum_m B^j_{lm}\xi_m = 0$ since: for $l = n$ the sum equals $\det B^j = 0$ and for $l \ne n$ it vanishes by an elementary linear algebra. The vector ξ is an eigenvector of B. It is a polynomial in the eigenvalues λ_j and in the elements of the matrix B. It vanishes at Λ_j.

5.3   Complete systems of Floquet solutions

Let us take any basis $\{\varphi_j \mid j \in \mathbb{Z}_0\}$ of the Hilbert scale $\{Z_s\}$ and assume that the basis is symplectic, i.e.,

$$\alpha_2[\varphi_j, \varphi_{-k}] = \langle \overline{J}\varphi_j, \varphi_{-k} \rangle = \frac{\delta_{j,k}}{\nu^J_j} \quad \text{for all } j, k \in \mathbb{Z}_0, \tag{5.5}$$

where $\nu^J_j = -\nu^J_{-j}$ and $\nu^J_j > 0$ if j is positive. Due to (5.5), $\overline{J}\varphi_k = \varphi_{-k}/\nu^J_k$ for any $k \in \mathbb{Z}_0$. Since $\overline{J}$ is an anti selfadjoint isomorphism of the scale $\{Z_s\}$ of order $-d_J \le 0$, then $C_1^{-1}j^{d_J} \le \nu^J_j \le C_1 j^{d_J}$ for every $j \ge 1$ with some $C_1 \ge 1$. For any real s we denote by Y_s the Hilbert space

$$Y_s = \overline{\operatorname{span}}\{\varphi_j \mid j \in \mathbb{Z}_n\} \subset Z_s.$$

The spaces $\{Y_s, \alpha_2 \mid_{Y_s}\}$ form a symplectic Hilbert scale with the basis $\{\varphi_j \mid j \in \mathbb{Z}_n\}$. In Y_0^c we choose some complex basis $\{\psi_j \mid j \in \mathbb{Z}_n\}$ and assume that it is symplectic:

$$\alpha_2[\psi_j, \psi_{-k}] = \delta_{j,k}\mu_j. \tag{5.6}$$

Besides, we assume that for big j it agrees with the basis $\{\varphi_j\}$:

$$\mu_j = i/\nu_j^J \quad \text{and} \quad \psi_{\pm j} = (\varphi_j \mp i\varphi_{-j})/\sqrt{2} \quad \text{if } j \geq j_1. \tag{5.7}$$

Example. If $\{\varphi_j\}$ is the trigonometric basis as in (1.1), i.e. $\varphi_k = \pi^{-1/2}\cos kx$ and $\varphi_{-k} = -\pi^{-1/2}\sin kx$, then for $|k| \geq j_1$ the functions ψ_k are complex exponents, $\psi_k = (2\pi)^{-1/2}e^{ikx}$.

Let $\{v_j\}$ be a system of Floquet solutions as in Section 5.2 and $\{\Psi_j\}$ are corresponding sections. For any $(r,\mathfrak{z}) \in R_0 \times \mathbb{T}^n$ we denote by $\Phi_1(r,\mathfrak{z})$ a linear map from $Y^c = Y_d^c$ to Z^c which identifies ψ_j with Ψ_j:

$$\Phi_1(r,\mathfrak{z}) : Y^c \to Z^c, \quad \psi_j \mapsto \Psi_j(r,\mathfrak{z}) \ \forall j \in \mathbb{Z}_n. \tag{5.8}$$

The map Φ_1 will be used to formulate an important assumption of completeness of a system of Floquet solutions. Before to do this we cut out of the set R_0 some "neighbourhood of infinity" and any neighbourhood of the singular set R_s to get an open domain R_1,

$$R_1 \Subset R_0 \setminus R_s.$$

Possibly, R_1 is disconnected. To simplify notations we assume that R_1 belongs to a single chart of the analytic manifold R_0 and treat R_1 as a bounded domain in $\mathbb{R}^n$. We fix any bounded complex domain R_1^c which contains R_1 with its δ-neighbourhood and does not intersect the singular set R_s^c. We denote by W_1 the set $W_1 = R_1 \times \mathbb{T}^n$ and denote by W_1^c its complex neighbourhood, $W_1^c = R_1^c \times \{|\operatorname{Im}\mathfrak{z}| < \delta\}$.

Definition 3. A system of Floquet solutions (5.4) which satisfies the analyticity assumptions a)–d) is called *complete* (in the space $Z = Z_d$) if:

0) it is formed by skew-orthogonal Floquet solutions,
and for any $(r,\mathfrak{z}) \in R_0 \times \mathbb{T}^n$ we have:

1a) the functions $\beta_j = i\alpha_2[\Psi_j(r,\mathfrak{z}), \Psi_{-j}(r,\mathfrak{z})]$, $j \in \mathbb{Z}_n$, are $\mathfrak{z}$-independent: $\beta_j = \beta_j(r)$,

b) there is a non-empty sub-domain of R where no function $\beta_j(r)$ vanishes identically,

c) the vectors $\{\Psi_j(r,\mathfrak{z})\}$ form a skew-orthogonal system in the space $T_{\Phi_0(r,\mathfrak{z})}^{\perp c} \mathcal{T}^{2n}$, that is:

$$\alpha_2[\Psi_j, \Psi_{-k}] = i\beta_j(r)\delta_{j,k} \quad \forall j, k. \tag{5.9}$$

2) The vectors $\{\Psi_j(r,\mathfrak{z})\}$ are uniformly asymptotically close to the complex basis $\{\psi_j\}$ and the exponents $\nu_j(r)$ are close to constants. Namely,

a) the linear map $\Phi_1(r,\mathfrak{z})$ equals to the natural embedding $\iota Y \hookrightarrow Z$ up to a Δ-smoothing operator, $\Delta > 0$:

$$\|\Phi_1(r,\mathfrak{z}) - \iota\|_{d,d+\Delta} \leq C_1 \quad \text{for all } (r,\mathfrak{z}) \in W_1^c; \tag{5.10}$$

b) for large j the functions $\beta_j(r)$ in (5.9) are analytic in R_1^c and are there close to the constants $(\nu_j^J)^{-1}$, defined in (5.5) (cf. (5.7)):

$$|\beta_j(r) - 1/\nu_j^J| \le C_2 |j|^{-\varkappa} \quad \text{for } r \in R_1^c \tag{5.11}$$

with some $\varkappa \ge \max(1, d_J + \Delta)$;

c) the exponents ν_j are bounded in R_1^c and are there "asymptotically close to constants". That is, $|\nu_j(r)| \le C_3 |j|^{d_A + d_J}$ for $r \in R_1^c$ and

$$|\nabla \nu_j(r)| \le C_4 |j|^{\tilde{\Delta}} \quad \text{for } r \in R_1^c \tag{5.12}$$

with some real $\tilde{\Delta} < d_A + d_J$.

The constants $C_1 - C_4$ in this definition may depend on the domain R_1 but not on j.

Since $\Psi_{-j}(w) = \overline{\Psi}_j(w)$ for real w and big j, then the corresponding functions β_j are real and $\beta_{-j} \equiv -\beta_j$. As $\mu_j \ge C^{-1} j^{-d_J}$, then by the assumption (5.11) we have that $|\beta_j(r)| \ge \frac{1}{2}\mu_j$ for all $r \in R_1^c$ and $j > j_2$, with some new constant j_2. We consider the product

$$\tilde{b}(r) = \prod_{j=n+1}^{j_2} \beta_j^2(r).$$

It follows from the properties c) and d) that the function $\tilde{b}$ is analytic in Π^c (see [KK]). Due to 1b) a set of its zeroes in R^c is a proper analytic subset. We add it to the complex singular set R_s^c and accordingly modify the sets R_s and R_0. If it is necessary, we also decrease the domain R_1 so that the inclusion $R_1 \Subset R_0 \setminus R_s$ still holds true.

Remark. The set R_s as it is defined now is the final singular set for our constructions. It comprises: 1) the singular part of the algebraic set R, 2) the set of degeneracy of the pull-back symplectic structure $\Phi_0^* \alpha_2$, 3) algebraic singularities of the Floquet exponents and 4) points where any two of them coincide. Finally, it contains 5) the zero-set of the function $\tilde{b}$ we have just constructed. The last set is a set of degeneracy of the system $\{\Psi_j\}$ since a vector $\Psi_j(r, \mathfrak{z})$ is skew-orthogonal to the tangent space $T_u \mathcal{T}^{2n}$ and to all the vectors $\Psi_k(r, \mathfrak{z})$ as soon as $\beta_j(r) = 0$. In the same time in Lemma 8 below we prove that the vectors $\{\Psi_j\}$ form a basis of the skew-orthogonal space $T_u^{\perp c} \mathcal{T}^{2n}$ if $r \notin R_s$.

The set $R_1 \Subset \tilde{R} \setminus R_s$ may be chosen to occupy most part of R_0 in the sense of measure: If $\tilde{R}$ is a bounded chart of the manifold R_0, mes_n is the Lebesgue measure in $\tilde{R}$ and γ is any positive number, then R_1 can be chosen in such a way that

$$\text{mes}_n(\tilde{R} \setminus R) \le \gamma. \tag{5.13}$$

A complete system of skew-orthogonal Floquet solutions spans the skew-orthogonal spaces $T_u^{\perp c} \mathcal{T}^{2n}$ in conformity with the term "complete" we use:

Lemma 8. *For any* $w \in W_1^c$ *the map* $\Phi_1(w)$ *defines an isomorphism of the spaces* Y^c *and* $T_u^{\perp c}\mathcal{T}^{2n}$, *where* $u = \Phi_0(w)$. [9]

Proof. By (5.10) the map $\Phi_1(w)$ is a compact perturbation of the embedding $\iota\, Y^c \to Z^c$, so $\mathrm{ind}_{\mathbb{C}}\, \Phi_1(w) = \mathrm{ind}\, \iota = 2n$. As a range of Φ_1 lies in $T_u^{\perp c}\mathcal{T}^{2n}$, then $\dim_{\mathbb{C}} \mathrm{Coker}\, \Phi_1 \geq 2n$. So if we can show that $\mathrm{Ker}\, \Phi_1 = \{0\}$, then the range of Φ_1 equals $T_u^{\perp c}\mathcal{T}^{2n}$ and the assertion will follow. Suppose that the kernel is non-trivial. Then it contains a nonzero vector $\xi = \sum y_j \psi_j$ and we have $0 = \Phi_1 \xi = \sum y_j \Psi_j(w)$. The skew-inner product of the right-hand side with any vector $\Psi_{-j}(w)$ equals $i\, y_j \beta_j(r)$ (see (5.9)). Thus, $y_j \equiv 0$ and $\xi = 0$. Contradiction. $\square$

Decreasing in a need the complex neighbourhood W_1^c of W_1 we easily get the following result:

Lemma 9. *For any* $s \in [-d - d_J - \Delta, d + \Delta]$ *the operator* $\Phi_1(w) : Y_s^c \to Z_s^c$ *analytically depends on* $w \in W_1^c$ *and is uniformly bounded. Moreover, for any* s *as above the map* $\Phi_1(w) - \iota\, Y_s^c \to Z_{s+\Delta}^c$ *is analytic in* $w \in W_1^c$ *as well.*

Example ((Birkhoff-integrable systems, see [K1, p. 401). and [Kap, BKM]]) Let Z be a space of sequences $\xi = (x_1, y_1; x_2, y_2; \dots)$, given some Hilbert norm and given the "usual" symplectic structure by means of the 2-form $J\, d\xi \wedge d\xi$, where $J(x_1, y_1; \dots) = (-y_1, x_1; \dots)$. We denote $p_j = (x_j^2 + y_j^2)/2$, $q_j = \mathrm{Arg}(x_j + iy_j)$ and consider an analytic hamiltonian $h(p_1, p_2, \dots)$. The subspace $\mathcal{T}^{2n} \subset Z$ formed by all vectors ξ such that $0 = x_{n+1} = y_{n+1} = \dots$ is invariant for the Hamiltonian vector field V_h and the restricted to $\mathcal{T}^{2n}$ system is obviously integrable. Let us denote $p^n = (p_1, \dots, p_n)$, $q^n(t) = (q_1, \dots, q_n)$ and denote by ν_j the functions $\nu_j(p^n) = \frac{\partial h(p^n, 0, \dots)}{\partial p_j}$, $j \geq 1$. We shall identify p^n with the vector $(p^n, 0, \dots)$.

The manifold $\mathcal{T}^{2n}$ is filled with solutions

$$\xi(t) = \{p^n = \mathrm{const}, q^n = t\nu^n(p^n) + \varphi^n; \ p_r = 0 \text{ for } r > n\},$$

where $\varphi^n \in \mathbb{T}^n$ and $\nu^n = (\nu_1, \dots, \nu_n)$. For any $j > n$ let us consider a smooth variation $\xi(t, \varepsilon)$ of a solution $\xi(t)$, which changes no action p_l except p_j and makes the latter equal ε^2:

$$\xi(t, \varepsilon) = \{p^n(t) = p^n, q^n(t) = t\nu^n(p^n) + q_0^n(\varepsilon); \ x_l(t) = y_l(t) = 0 \text{ if } l > n, l \neq j\}$$

and $x_j(t) = \varepsilon \cos(t\nu_j(p^n) + \varphi(\varepsilon))$, $y_j = \varepsilon \sin(t\nu_j(p^n) + \varphi(\varepsilon))$, where $\varphi(\varepsilon) \in S^1$. The curve $\tilde{v}_j = \frac{\partial}{\partial \varepsilon} \xi(t, \varepsilon)\big|_{\varepsilon=0}$ is a solution of the equation, linearised about $\xi(t)$. It equals

$$\tilde{v}_j(t, \varphi) = \{\delta p^n = 0, \delta q^n = q_0^{n\prime}(0); \ \delta x(t), \delta y(t)\},$$

---

[9] For a complex w and $u = \Phi_0(w)$ we define $T_u^{\perp c}\mathcal{T}^{2n}$ as the set of all $z \in Z^c$ such that $\alpha_2[z, \Phi_0(w)_*\xi] = 0$ for any $\xi \in T_w W_1^c$.

where $\delta x_r(t) = \delta y_r(t) = 0$ if $r \neq j$ and

$$\delta x_j = \cos(t\nu_j(p^n) + \varphi), \quad \delta y_j = \sin(t\nu_j(p^n) + \varphi), \quad \varphi = \varphi(0).$$

The curve $v_{triv}(t) = \{\delta p^n = 0, \delta q^n = q_0^{n\prime}(0); \ \delta x = \delta y = 0\}$ is a trivial solution of the linearised equation (it may be obtained using a variation of $\xi(t)$, corresponding to a rotation of the phase-vector q^n). An appropriate complex linear combination of the solutions $\tilde{v}_j(t, 0)$, $\tilde{v}_j(t, \pi)$ and of the trivial solution as above takes the Floquet form $v_j(t) = e^{i\nu_j(p^n)t}\Psi_j$, where $\Psi_j = (0, \ldots; i, 1; 0, \ldots)$ (the pair $(i, 1)$ stands on the jth place).

Let us suppose that $|\nu_j| \leq Cj^{d_A}$ for some d_A and that (5.12) holds. Then the system of Floquet solutions $\{v_j, \overline{v}_j \mid j \geq n+1\}$ is complete in the sense of Definition 3.

This example illustrates well the definition but it is too simple and so too restrictive: to be Birkhoff integrable a finite-dimensional system has to have dim $Z/2$ integrals of motion, but to have a complete system of Floquet solutions for the equations linearised about solutions in $\mathcal{T}^{2n}$ it needs only n of them (see below Proposition 2).

To be useful in analytical studies of the equation (5.1) and its perturbations, a system of Floquet solutions should be complete and *non-resonant*:

Definition 4. A system of Floquet exponents $\{\nu_j(r) \mid j \in \mathbb{Z}_n\}$ satisfying a)–c) is called *non-resonant* if:

3) there exists a domain $O \subset R_0$ such that for all $s \in \mathbb{Z}^n$ and all $j, k \in \mathbb{Z}_n$, $j \neq -k$, we have:

$$\omega(r) \cdot s + \nu_j(r) \not\equiv 0 \quad \text{in } O, \tag{5.14}$$

$$\omega(r) \cdot s + \nu_j(r) + \nu_k(r) \not\equiv 0 \quad \text{in } O. \tag{5.15}$$

The system of Floquet solutions with non-resonant exponents also is called non-resonant.

Since the exponents ν_j are algebraic (or analytic) functions, then zero-set of any resonance is nowhere dense.

Finally we give:

Definition 5. A system of Floquet solutions (5.4) satisfying a)–d) is called complete non-resonant if it satisfies assumptions 0)–3) from Definitions 3, 4.

It turns out that the assumption 1) follows from 3):

Lemma 10. *Any non-resonant system of Floquet solutions satisfy assumptions 0),1a) and 1c) from Definition 3.*

Proof. To check 1c) we should prove that for any $j \neq -k$ the function $F(r, \mathfrak{z}) = \alpha_2[\Psi_j, \Psi_k]$ vanishes identically. To do it let us consider the function $f(t; r, \mathfrak{z})$,

$$f := e^{i(\nu_j + \nu_k)t}\alpha_2[\Psi_j(w(t)), \Psi_k(w(t))] = \alpha_2[v_j(t), v_k(t)],$$

where $w(t) = (r, \mathfrak{z} + t\omega(r))$. Since the flow-maps $S^{t_2}_{t_1*}$ are symplectic [KK], then the skew-product of any Floquet solutions v_j and v_k is time-independent and

$$0 = \frac{df}{dt}\bigg|_{t=0} = i(\nu_j + \nu_k)F + \nabla_{\mathfrak{z}} F \cdot \omega.$$

Let us expand F as Fourier series, $F = \sum e^{is \cdot \mathfrak{z}} \widehat{F}(r, s)$. From the last identity we get that $\widehat{F}(r, s)(\nu_j + \nu_k + s \cdot \omega(r)) = 0$ for all s and all r. Since the second factor is nonzero for almost all r, so $\widehat{F}(r, s) \equiv 0$ and $F(r, \mathfrak{z}) \equiv 0$.

If $j = -k$, then we have: const $\equiv \alpha_2[v_j(t), v_{-j}(t)] = \alpha_2[\Psi_j(w(t)), \Psi_{-j}(w(t))]$. Because the assumption iv) (Section 3), the curve $w(t)$ is dense in a torus $\{r\} \times \mathbb{T}^n$ for almost all r. So the left-hand side of (5.9) with $k = -j$ is $\mathfrak{z}$-independent for almost all r. By continuity, it is $\mathfrak{z}$-independent for all r, as states 1a).

To check 0) one has to argue similar, using variations δr, $\delta \mathfrak{z}$ of the initial conditions for the curve $w(t)$. $\square$

Corollary. *A system of Floquet solutions* (5.4) *which meets the assumptions* a)–d) *from Section 5 as well as the assumptions* 2), 3) *from Definitions 3, 4 is skew-orthogonal to* $\mathcal{T}^{2n}$ *and is complete non-resonant, provided the assumption* 1b) *holds. The latter happens e.g., if there exists a point* $r_* \in \overline{R}$ *such that* $\Psi_j(r, \mathfrak{z}) \to \psi_j$ *as* r *tends to* r_**. Here* $\overline{R}$ *signifies the closure of* R *in* $\mathbb{R}^N$ *where* R *is a subset.*

Practically the point r_* corresponds to the zero-solution of the equation (5.1) (or another trivial solution).

This result simplifies verification of completeness for a system of Floquet solutions since it is much easier to check the non-resonance relations (5.14), (5.15) than the completeness 1a)–1c).

The transformation Φ_1 integrates the linearised equation (5.2): it sends the curves $y_j = e^{i\nu_j(r_0)t}\psi_j$ to solutions $v_j(t)$ of (5.2). It is convenient to have this transformations symplectic. For this end the sections $\{\Psi_j\}$ have to be properly reordered and normalised by multiplying by some analytic functions of r; simultaneously the basis $\{\psi_j\}$ also have to be transformed by a linear symplectomorphism which changes finitely many its components only. In this way the following result can be proven:

Proposition 2. *Given any complete system of Floquet solutions* (5.4) *we can normalise the sections* $\{\Psi_j\}$ *and the basis* $\{\psi_j\}$ *is such a way that the new basis still meets* (5.6), (5.7) *and the new system of Floquet solutions still is complete. Besides,*

a) *for any* $(r, \mathfrak{z}) \in W_1 = R_1 \times \mathbb{T}^n$ *the map* $\Phi_1(r, \mathfrak{z})$ *defines a symplectic isomorphism of* Y *and* $T^{\perp}_{\Phi(r, \mathfrak{z})} \mathcal{T}^{2n}$ *which analytically in* $(r, \mathfrak{z}) \in W_1^c$ *extends to a bounded linear map* $Y^c \to Z^c$*;*

b) *the nonautonomous linear map* $\Phi_1(r, \mathfrak{z} + t\omega(r))$ *sends solutions of an autonomous equation* $\dot{y} = JB(r)y$*,* $y \in Y^c$*, to solutions of* (5.2). *The selfadjoint* *operator $B(r)$ is analytic in r. Besides, $\operatorname{ord} B(r) \leq d_A$ and $\operatorname{ord} \nabla_r B(r) \leq -d_J - \tilde{\Delta}$. Spectrum of the operator $JB(r)$ equals to the set of Floquet exponents of the solutions (5.4).*

· The basis $\{\psi_j\}$ may depend on a connected component of the set R_1.

The *leading Lyapunov exponent* of equation (5.2) in Z_d is a number a equal to the supremum over all real numbers a' such that $\overline{\lim}_{t \to \infty} e^{-a't} \|v(t)\|_d = \infty$ for some solution $v(t) \subset Z_d$ of (5.2). A solution $u_0(t)$ of (5.1) is called *linearly stable* if the leading Lyapunov exponent of the corresponding linearised equation (5.2) vanishes.

A direct consequence of Proposition 2 is the following

Corollary. *If the linearised equations* (5.2) *have complete system of Floquet solutions, then the leading Lyapunov exponent of the equation corresponding to a solution $u_0 = u_0(t; r, з)$ with $r \in R_1$ equals $\nu^I(r) = \max\{\operatorname{Im} \nu_j(r) \mid n < |j| < j_1\}$.*[10]

Indeed, by the proposition any variation $u'(t)$ of a solution $u_0(t)$ can be written as $\Phi_0(u_0)_*(r', з') + \Phi_1(u_0)y'$ and in terms of the prime-variables the equation (5.2) reads as

$$\dot{r}' = 0, \quad \dot{з}' = \omega(r)_* r', \quad \dot{y}' = JB(r)y'. \tag{5.16}$$

Decomposing $y'(0)$ in the basis $\{\psi_j\}$ we find that $e^{-at} \|u'(t)\|_s \to 0$ as t grows, if $a > \nu^I(r)$. If $a < \nu^I(r)$ and ψ_j is an eigenvector of $JB(r)$ with the eigenvalue ν_j such that $\operatorname{Im} \nu_j = a$, then $y'(t) = e^{-i\nu_j t} \psi_{-j}$ is the y'-component of a solution of (5.16). A norm of this solution grow with t faster than e^{at}.

5.4 Lower-dimensional invariant tori of finite-dimensional systems and Floquet's theorem

Let O be a domain in the Euclidean space $\mathbb{R}^{2N}$, given the usual symplectic structure. Let $H_1, \ldots, H_n$, $1 \leq n < N$, be a system of commuting hamiltonians, defined and analytic in O. Let $T^n \subset O$ be a torus, analytically embedded in O, which is invariant for all n Hamiltonian vector fields V_{H_j}. The vector fields are assumed to be linearly independent at any point of the torus.

Under mild nondegeneracy assumptions on the system of hamiltonians (see [Nek]), the torus T^n can be proven to belong to an n-dimensional family of invariant n-tori T_r^n:

$$T^n \subset \mathcal{T}^{2n} = \bigcup_{r \in R} T_r^n, \quad 0 \in R \Subset \mathbb{R}^n; \quad T^n = T_0^n,$$

where $\mathcal{T}^{2n}$ is an analytic $2n$-dimensional submanifold of O. Moreover, the symplectic form, restricted to $\mathcal{T}^{2n}$, is nondegenerate and $\mathcal{T}^{2n}$ admits analytic

---

[10] We recall that the functions $\nu_j(r)$ with $|j| \geq j_1$ are real valued by the assumption b).

coordinates $(r, \mathfrak{z})$, $\mathfrak{z} \in \mathbb{T}^n$, such that for every $j = 1, \ldots, n$ the vector field V_{H_j}, restricted to $\mathcal{T}^{2n}$, takes the form $\sum_l \omega_j^l(r) \partial/\partial_{\mathfrak{z}l}$ (the functions $\omega_j^l(r)$ all are analytic).

Instead of presenting here the nondegeneracy assumptions, we just *assume* existence of a family of invariant n-tori as above. Then for any r there exist linear combinations $K_1, \ldots, K_n$ of the original hamiltonians H_j such that for every j the vector field V_{K_j} restricted to the torus T_r^n equals $\partial/\partial_{\mathfrak{z}j}$. Accordingly, at any point $(r, \mathfrak{z}) \in T_r^n$ every vector field V_{K_j} defines $N - n$ Floquet multipliers $e^{i\lambda_l^j(r)}$, $l = 1, \ldots, N - n$, corresponding to directions, transversal to $\mathcal{T}^{2n}$. [11] For simplicity we assume that $\mathcal{T}^{2n}$ is a linearly stable invariant set of every vector field V_{K_j} (so also of every V_{H_j}). Then all the functions $\lambda_l^j(r)$ are real.

The following result is a version of the Floquet theorem "for multidimensional time". For a proof see [K3] and [KK].

Proposition 3. *Under the given above assumptions, every vector field V_{H_j}, linearised about its solutions in $\mathcal{T}^{2n}$, has a complete system of $N - n$ skew-orthogonal Floquet solutions with real exponents $\nu_j(r)$.*

We note that in the finite-dimensional situation which we discuss now, the item 2) of Definition 3 becomes trivial.

6   Linearised Lax-integrable equations

Now we pass to the problem of constructing complete systems of Floquet solutions of infinite-dimensional sustems. If (3.1) was a finite-dimensional system (i.e., dim $Z < \infty$) with an integrable subsystem $\mathcal{T}^{2n}$ as above, then by Proposition 2 linearised equation (5.2) would have a complete system of Floquet solutions provided that the equation (3.1) had n nondegenerate integrals in involution. For infinite-dimensional systems the Floquet theorem is unknown. In this section we show that for Lax-integrable equations Floquet solutions can be constructed as quadratic forms of the eigen-functions of the corresponding $\mathcal{L}$-operator.

6.1   Abstract situation

Let $u(t)$ be any smooth solution of a Lax-integrable equation (5.1)=(2.6). For any smooth vector v we denote $\mathcal{L}_t'(v) = \mathcal{L}_{u(t)}'(v) = \frac{\partial}{\partial \varepsilon} \mathcal{L}_{u(t)+\varepsilon v}\big|_{\varepsilon=0}$, and

---

[11] The multipliers are defined as eigenvalues of the linearized time-2π flow-map of the vector field V_{K_j}, restricted to a skew-orthogonal component to the space $T_{(r,\mathfrak{z})}\mathcal{T}^{2n}$. They are $\mathfrak{z}$-independent, see [K3].

similar define operators $\mathcal{A}'_t(v)$. Differentiating equation (3.4) in a direction v, we get a Lax-representation for the linearised equation (5.2):

$$\frac{d}{dt}\mathcal{L}'_t(v) = [\mathcal{A}'_t(v), \mathcal{L}_t] + [\mathcal{A}_t, \mathcal{L}'_t(v)],$$

where $\mathcal{A}_t = \mathcal{A}_{u(t)}$ and $\mathcal{L}_t = \mathcal{L}_{u(t)}$. Let us consider smooth eigenvectors of the operator $\mathcal{L}_0 = \mathcal{L}_{u_0}$ and of its conjugate operator $\mathcal{L}_0^*$, corresponding to the same eigenvalue λ:

$$\mathcal{L}_0\chi_0 = \lambda\chi_0, \qquad \mathcal{L}_0^*\xi_0 = \lambda\xi_0.$$

We assume that the following initial-value problems,

$$\dot{\chi}(t) = \mathcal{A}_t\chi(t), \quad \chi(0) = \chi_0, \qquad \dot{\xi}(t) = -\mathcal{A}_t^*\xi(t), \quad \xi(0) = \xi_0, \qquad (6.1)$$

have smooth solutions $\chi(t)$ and $\xi(t)$. Then for any t we have $\mathcal{L}_t\chi(t) = \lambda\chi(t)$ and $\mathcal{L}_t^*\xi(t) = \lambda\xi(t)$ (see Lemma 2.3 for the proof of the first relation; proof of the second is identical).

We claim that

$$\frac{d}{dt}\langle \mathcal{L}'_t(v(t))\chi, \xi \rangle = 0. \qquad (6.2)$$

Indeed, abbreviating $\mathcal{L}'_t(v(t))$ to $\mathcal{L}'$ and $\mathcal{A}'_t(v(t))$ to $\mathcal{A}'$, we write the left-hand side of (6.2) as

$$\langle \mathcal{L}'\chi, \dot{\xi} \rangle + \langle \dot{\mathcal{L}}'\chi, \xi \rangle + \langle \mathcal{L}'\dot{\chi}, \xi \rangle$$
$$= \langle \mathcal{L}'\chi, -\mathcal{A}^*\xi \rangle + \langle ([\mathcal{A}', \mathcal{L}] + [\mathcal{A}, \mathcal{L}'])\chi, \xi \rangle + \langle \mathcal{L}'\mathcal{A}\chi, \xi \rangle$$
$$= \langle [\mathcal{A}', \mathcal{L}]\chi, \xi \rangle = \langle \mathcal{A}'\mathcal{L}\chi, \xi \rangle - \langle \mathcal{A}'\chi, \mathcal{L}^*\xi \rangle = (\lambda - \lambda)\langle \mathcal{A}'\chi, \xi \rangle = 0.$$

Since $\mathcal{L}'_t(w)$ linearly depends on $w \in Z_{s'}$ as an operator from $\mathfrak{Z}_{s'}$ to $\mathfrak{Z}_{s'-d}$, then

$$\langle \mathcal{L}'_t(w)\chi, \xi \rangle_3 = \langle w, q_t(\chi, \xi) \rangle_Z \quad \forall w, \qquad (6.3)$$

where $q_t(\chi, \xi) = q_{u(t)}(\chi, \xi)$ is an $Z_{-s'}$-valued quadratic form of $\chi, \xi \in \mathfrak{Z}_{s'}$, which is C^1-smooth in t. Hence, we can rewrite (6.2) as

$$\frac{d}{dt}\langle v(t), q_t(\chi, \xi) \rangle \equiv 0. \qquad (6.4)$$

For a moment let us denote $q_t(\chi, \xi) = w$. Then

$$\langle v, A_t J w \rangle = -\langle J A_t v, w \rangle = -\langle \dot{v}, w \rangle = \langle v, \dot{w} \rangle, \qquad (6.5)$$

where the last equality follows from (6.4). At this point we assume that the flow-maps $S^t_{\tau*}(u(\tau))$ of the linearised equation (5.2) preserves the space Z_∞. Then the set $\{v(t)\}$ formed by values at time t of all smooth solutions of equation (5.2) equals Z_∞, so $\dot{w} = A(t)Jw$ since (6.5) holds for any t and for all solutions $v(\cdot)$. Therefore $J\dot{w} = JA(t)Jw$, i.e. the curve $Jw(t) = J(q(\chi(t), \xi(t)))$ satisfies the equation (5.2).

Thus, linearised Lax-integrable equations have solutions which can be obtained as bilinear forms of eigen-functions of the $\mathcal{L}$-operator and its adjoint:

Theorem 3. *If flow-maps of the linearised equation* (5.2) *preserve the space* Z_∞ *and the curves* $\chi(t)$, $\xi(t)$ *are smooth solutions of equations* (6.1), *then the function* $J\big(q_t(\chi(t), \xi(t))\big)$ *with* q_t *defined in* (6.3) *solves the linearised equation* (5.2).

Remark. Let $\{Z_s\}$ be a Sobolev scale of 2π-periodic functions of a space-variable x. If the flows of linear equations (5.2) and (6.1) preserve Sobolev spaces of 2π-antiperiodic functions, the curves $\chi(t), \xi(t)$ are constructed as above and (6.3) holds with some 2π-periodic function $q_t(\chi, \xi)(x)$, then due to the same arguments as before the function $J(q)$ solves (5.2).

Remarkably the solutions given by the theorem and the remark have the Floquet form (5.3) and jointly form a complete non-degenerate family. Below we check this property for the KdV and SG equations.

6.2   Linearised KdV equation

Now we consider the KdV equation and take for the invariant manifold $\mathcal{T}^{2n}$ a bounded part of any finite-gap manifold $\mathcal{T}_V^{2n}$ of the form (4.5). We have already checked that it satisfies assumptions i)–iv) from Section 5.

For any n-gap solution $u_0(t, \cdot) \in T_V^n(r)$ the equation linearised about u_0 takes the form

$$\dot{v} = \frac{1}{4}\, v_{xxx} + \frac{3}{2}\, \frac{\partial}{\partial x}\, (u_0(t, x)v). \tag{6.6}$$

Since $u_0(t, x)$ is a smooth function, then this equation is well-defined in Sobolev spaces H_0^d with $d \geq 1$. Thus the assumption v) on the invariant manifold also is satisfied.

The equation (6.6) has trivial solutions, obtained by differentiations in directions, tangent to the n-gap tori. They can be written as $\frac{\partial u(t,x;r,\jmath)}{\partial \jmath_j}$ (see (4.4)).

We recall that the $\mathcal{L}$-operator of the KdV equation is the Sturm–Liouville operator $\mathcal{L} = -\partial^2/\partial x^2 - u_0(t, x)$ and consider any its complex eigenfunction $\chi(x; \lambda)$ with an eigenvalue λ, satisfying the Floquet–Bloch boundary conditions:

$$\mathcal{L}\chi(x; \lambda) = \lambda\chi(x; \lambda), \qquad \chi(x + 2\pi; \lambda) = e^{i\rho}\chi(x; \lambda), \ \ \rho = \rho(\lambda). \tag{6.7}$$

One of the most important and elegant properties of the KdV equation (and of the whole class of Lax-integrable equations) is that χ as a function of λ is meromorphic in $\Gamma \setminus \infty$ ($\Gamma = \Gamma(r)$ is the Riemann surface defined in Section 4) and can be normalised to have at infinity the singularity $\exp i\sqrt{\lambda}\, x$. This function admits a representation in terms of the same theta-function θ and the vectors V, W, $\jmath$ as in Section 4. The representation is given by the

Its-Matveev formula (see in [DMN,D,BB]):

$$\chi(t,x;r,\mathfrak{z};P) = e^{\Omega_1(P)x+\Omega_3(P)t}\,\frac{\theta(A(P)+i(Vx+Wt+\mathfrak{z}))\theta(i\mathfrak{z})}{\theta(A(P)+i\mathfrak{z})\theta(i(Vx+Wt+\mathfrak{z}))},$$
$$P = (\lambda,\mu) \in \Gamma.$$

Here $A(P)$ is the Abel transformation, the same as above, and Ω_1, Ω_3 are Abel integrals of the differentials $d\Omega_1$, $d\Omega_3$. The integrals are defined modulo periods of the differentials. For $P = (\lambda,\mu)$ with real λ we normalise the integrals in the following way:

$$\Omega_j(\lambda,\mu) = \int_{E_1}^{\lambda} d\Omega_j \quad \text{for } (\lambda,\mu) \in \Gamma_+,\ \lambda \in \mathbb{R},$$

where $[E_1,\lambda]$ stands for the path in Γ_+ through upper edges of the cuts. We denote by σ the holomorphic involution of Γ which transposes the sheets, $\sigma(\lambda,\mu) = (\lambda,-\mu)$. Denoting for any $P = (\lambda,\mu) \in \Gamma$ by γ_P the path from $\sigma(P)$ to P through E_1, equal to $\gamma_P = \sigma(-[E_1,\lambda]) \cup [E_1,\lambda]$, we get that

$$\Omega_j(P) = \frac{1}{2}\int_{\gamma_P} d\Omega_j,$$

since $\sigma^*d\Omega_j = -d\Omega_j$ due to the normalisation (4.2). Now we take a point P close to infinity and denote by μ_P the path from $\sigma(P)$ to P equal to a lift to Γ of the circle in $\mathbb{C}_\lambda$ centred at infinity, which passes through λ and is cut there. The loop $\gamma_P - \mu_P$ is contractible in Γ since it envelops all the cuts, so $\Omega_j(P) = \frac{1}{2}\int_{\mu_P} d\Omega_j$. Using this equality and (4.2) (with $c = 0$) we get the asymptotics:

$$\Omega_1(P) = k + O(k^{-2}), \quad \Omega_3(P) = k^3 + O(k^{-1}), \quad k = i\sqrt{\lambda}. \tag{6.8}$$

Let us denote

$$f(U;r,\mathfrak{z};P) = \frac{\theta(A(P)+iU+i\mathfrak{z})\theta(i\mathfrak{z})}{\theta(A(P)+i\mathfrak{z})\theta(iU+i\mathfrak{z})} \tag{6.9}$$

and rewrite χ as

$$\chi(t,x;r,\mathfrak{z};P) = e^{\Omega_1(P)x+\Omega_3(P)t}f(Vx+Wt;r,\mathfrak{z};P). \tag{6.10}$$

By the Riemann theorem (see [D,BB]) the first term of the denominator in the right-hand side of (6.9) has exactly n zeroes which form poles of the function $P \mapsto f$ and lie in the ovals $a_1,\ldots,a_n$. Since $|\theta(iU+i\mathfrak{z})| \geq C(r) > 0$ and $|\theta(A(\infty)+i\mathfrak{z})| \geq C(r)$, then the function $f(U;r,\mathfrak{z};P)$ is analytic and bounded for r from an appropriate complex neighbourhood of any compact subset of R and for $(\lambda, \mathfrak{z}, U)$ from the complex domain

$$\{\, |\mathrm{Im}\,\lambda|, |\mathrm{Im}\,\mathfrak{z}|, |\mathrm{Im}\,U| < \delta, \ \ \mathrm{Re}\,\lambda > E_1 \ \ \text{and} \ \ \mathrm{dist}\,(\lambda, \Delta_{V_j}) > \delta \ \forall j\}, \quad (6.11)$$

where $\delta > 0$ is sufficiently small.

We recall that the closed gaps $[\lambda_{2j-1} = \lambda_{2j}]$ are labelled by indices $j \in \mathbb{N}_V$. So for any $P = P_{\pm j}$, where $P_{\pm j} = \{\pm\sqrt{R(\lambda_{2j})}, \lambda_{2j}\} \in \Gamma$, the function $\chi(t, x; P_{\pm j})$ must be 4π-periodic in x. Since f is 2π-periodic, we should have $\Omega_1(P_{\pm j}) \in \frac{i}{2}\mathbb{Z}$. This relation holds identically in r. When r tends to zero, λ_{2j} tends to $j^2/4$ and $\Omega_1(P_j)$ tends to $ij/2$ (this follows from elementary calculations, see [KK]). Therefore,

$$\Omega_1(P_j) = \frac{i}{2}\,j, \qquad j \in \mathbb{N}_V. \qquad (6.12)$$

Conversely, for any P which meets (6.12) the function (6.10) is 4π-periodic.

Since the operator $\mathcal{A}$ for the KdV equation is anti selfadjoint, then the second equation in (6.1) coincides with the first and we can take $\xi(t) = \chi(t)$. Now the quadratic form q as in Theorem 3 equals χ^2. Finally, since $J = \partial/\partial x$, then the solutions of the linearised equation (5.2)=(6.6) constructed in Theorem 3 are curves $v_j(t) \in Z$ of the form

$$v_j(t, x; r, \mathfrak{z}) = \left(\frac{(2\pi)^{-1/2}}{2\Omega_1(P_j)}\right) \frac{\partial}{\partial x}\left(e^{2(\Omega_1(P_j)x + \Omega_3(P_j)t)} f^2(Vx + Wt; r, \mathfrak{z}; P_j)\right). \tag{6.13}$$

Here $j \in \mathbb{Z}_V$, $P_j = P_j(r)$ and the first factor in the right-hand side is a convenient normalisation. Thus we have obtained a system of Floquet solutions of the form (5.4),[12] where the sections Ψ_j of the bundle $T^c H_0^d|_{\mathcal{T}^{2n}}$ have the form

$$\Psi_j(r, \mathfrak{z})(x) = \frac{\partial}{\partial x}\left(\frac{e^{2\Omega_1(P_j)x}}{2\sqrt{2\pi}\,\Omega_1(P_j)}\, f^2(Vx; r, \mathfrak{z}; P_j)\right), \quad j \in \mathbb{Z}_V, \qquad (6.14)$$

and the exponents ν_j are $\nu_j(r) = -2i\Omega_3(P_j) = -2i\int_{E_1}^{P_j} d\Omega_3$.

Since the differential $i\,d\Omega_3$ is real (see (4.2′)), then the exponents $\nu_j(r)$ are real for real r and are analytic in r (they have no algebraic singularities). We claim that this system is complete non-resonant. To simplify notation we suppose that $V = (1, \ldots, n)$. Now the complex basis $\{\psi_j \mid j \in \mathbb{Z}_0\}$ is the exponential basis $\psi_j = e^{ijx}/\sqrt{2\pi}$.

The system of Floquet exponents is non-resonant. To prove the non-resonance we may assume that the vector r is sufficiently small. For any

---

[12] the set of indices $\mathbb{Z}_V$ which we use now is in obvious 1-1 correspondence with the set $\mathbb{Z}_n$.

$j \in \mathbb{Z}_V = \mathbb{Z}_n$ we denote by $V^{(n+1)}$ the $(n+1)$-vector (V, j) and view the torus $T_V^n(r)$ as a degenerate $(n+1)$-gap torus $T_{V^{(n+1)}}^{n+1}(r, 0)$. It can be checked [KK] that the integral which defines $\nu_j(r)$ equals $W_{n+1}^{(n+1)}(r, 0)$. Since the frequency vector ω for finite-gap solutions which fill the torus $T_V^n(r)$ is $\omega = W$, then the non-resonance relation (5.18) which has to be checked takes the form

$$\sum_{l=1}^{n} W_l^{(n+1)}(r, 0)s_l + W_{n+1}^{(n+1)}(r, 0) \neq 0. \tag{6.15}$$

We can suppose that $s \neq 0$; say, $s_1 \neq 0$. By Lemma 7, for $r = (\varepsilon, 0, \dots, 0)$ we have: $W_l^{(n+1)} = \text{const} + \delta_{l,1} \frac{3}{8V_1} \varepsilon^2 + O(\varepsilon^4)$. Therefore, the left-hand side of (6.15) equals $\text{const} + s_1 \frac{3}{8V_1} \varepsilon^2 + O(\varepsilon^4)$. It does not vanish identically and (6.15) follows.

The system is complete. By the Corollary to Lemma 10 we should only check the assumptions 1b) and 2) from Definition 3. The function $f(\cdot; r, \mathfrak{z}; P)$ converges to unit as $r \to 0$ (as well as the theta-function, see [KK]). Therefore $\Psi_j(x)$ converges to the complex exponent $(2\pi)^{-1/2}e^{ijx} = \psi_j(x)$, so 1b) follows and it remains to check the item 2).

Given any $\gamma > 0$ we fix a subset $R_1 \Subset R$ such that $\text{mes}(R \setminus R_1) < \gamma$ (see (5.13)). For $r \in R_1$ we shall check the properties 2a)–2c).

First we show that the map Φ_1 defined in (5.8) is close to the embedding ι. Since $\Psi_{-j} = \overline{\Psi}_j$, we have to examine the vectors Ψ_j with $j \in \mathbb{N}_V$ only. Since the potentials $u(x)$ have zero mean-values, then $\lambda(P_j) = \frac{1}{4}j^2 + O(j^{-1})$ (see [Ma,MT]) and $k(P_j) = i\sqrt{\lambda} = \frac{i}{2}j + O(j^{-2})$. Using (6.8) we get that $\Omega_3(P_j) = -\frac{i}{8}j^3 + O(j^{-1})$. So

$$\nu_j(r) = -\tfrac{1}{4}j^3 + O(j^{-1}), \tag{6.16}$$

uniformly in r from a complex neighbourhood $R_1 + \delta$ of R_1.

Since the holomorphic differentials $d\omega_l$ have the form $d\omega_l = C(\lambda^{-3/2} + \dots)d\lambda$ (see [S, KK]), then

$$|A(P_j)| \leq C_1 \int_{j^2/4}^{\infty} \lambda^{-3/2}\, d\lambda \leq C_1|j|^{-1} \quad \text{uniformly in } r \in R_1 + \delta. \tag{6.17}$$

Therefore for all $P, U, \mathfrak{z}$ as in (6.11) and for r from $R_1 + \delta$ the function f is close to one:

$$|f(U; r, \mathfrak{z}; p) - 1| \leq C|j|^{-1}. \tag{6.18}$$

Using (6.12), (6.18) and the Cauchy estimate we find that the functions $\Psi_j(r, \mathfrak{z})$ defined in (6.14) are close to complex exponents:

$$\Psi_j(r, \mathfrak{z})(x) = \frac{1}{2\pi} e^{ijx}(1 + \zeta_j(r, \mathfrak{z})(x)), \tag{6.19}$$

where $|\zeta_j(r,\jmath)(x)| \le Cj^{-1}$ for $r \in R_1 + \delta \subset \mathbb{C}^n$, $|\operatorname{Im}\jmath| \le \delta$ and $|\operatorname{Im} x| \le \delta$ with some j-independent δ and $C = C(\delta)$.

To check the property 2a) from Definition 3 with $\Delta = 1$ we shall show that the map

$$\sum a_j e^{ijx} \mapsto \sum a_j \zeta_j(x)$$

is 1-smoothing, i.e., it sends a space $H_0^r(S^1)$ to the space $H_0^{r+1}(S^1)$. To do it we observe that in the Hilbert bases $\{(\sqrt{2\pi}\,j^r)^{-1}e^{ijx}\}$, $\{(\sqrt{2\pi}\,j^{r+1})^{-1}e^{ijx}\}$ of the two spaces the map has the matrix M with the entries $M_{lj} = l^{r+1}j^{-r}\int e^{i(j-l)x}\zeta_j(x)\,dx$. Since for $|\operatorname{Im} x| < \delta$ the function ζ_j is analytic and bounded by Cj^{-1}, then $|m_{lj}| \le C_\delta(l/j)^{r+1}e^{-\delta|j-l|}$. Therefore the l_1-norm of any row and any column of the matrix M is bounded by a constant C'. Hence, a norm of the map (6.19) as a map from H_0^r to H_0^{r+1} is bounded by the same C' due to the Schur criterion[13] and 2a) follows.

The property 2b) with $\varkappa = 3$ easily follows from (6.19).

The property 2c) with $d_A + d_J = 3$ and $\widetilde{\Delta} = 1$ is an immediate consequence of the asymptotic (6.16).

The system satisfies a)–d). [14] The first assertion of a) follows from the convergence $\nu_j(r) = -2i\Omega_3(P_j) \to -\frac{1}{4}j^3$ as $r \to 0$, which implies that for small r all the functions ν_j are distinct. The second assertion follows from the item 2c) of Definition 3 which is checked already with $d_A + d_J = 3$ and $\widetilde{\Delta} = 1$.

The assumption b) follows from 2a), 2c) since the exponents ν_j are real (for real r). The assumptions c), d) are now empty since all the Floquet exponents are analytic functions.

Finally for the domain R as in (4.5) we have proved the following result:

Theorem 4. *For any $\gamma > 0$ there exists a subset $R_1 \Subset R$, $\operatorname{mes}(R \setminus R_1) < \gamma$, such that on $\mathcal{T}_1^{2n} = \Phi_0(R_1 \times \mathbb{T}^n) \subset \mathcal{T}^{2n}$ the system of Floquet solutions (6.14) with $j \in \mathbb{Z}_V$ is complete non-resonant (in any space H_0^d, $d \ge 1$).*

6.3 Higher KdV-equations

The lth equation from the KdV-hierarchy has an $[L, A]$-pair with the same $\mathcal{L}$-operator $L = -\partial^2/\partial x^2 - u$ and with some $\mathcal{A}$-operator of the form $\mathcal{A} = \mathcal{A}_l = \operatorname{const} \partial^{2l+1}/\partial x^{2l+1} + \ldots$ (see [DMN,MT,ZM]). Solutions χ^l of equation (6.1) with $\mathcal{A} = \mathcal{A}_l$ are given by the Its–Matveev formula (6.6), where the differential Ω_3 should be replaced by an appropriate differential Ω_{2l+1} and

---

[13] The criterion states that a norm of a linear operator which in Hilbert bases of the two Hilbert spaces has a matrix $\{M_{lj}\}$, is no bigger than $(\sup_l \sum_j |M_{lj}| \cdot \sup_j \sum_l |M_{lj}|)^{1/2}$. See [HS] or [KK].

[14] see (5.4) and below.

the vector W – by the vector W^l, formed by b-periods of Ω_{2l+1}. We get Floquet solutions v_j^l of the linearised lth equation,

$$v_j^l(t, x; r, \mathfrak{z}) = e^{i\nu_j^l(r)t}\Psi_j(r, \mathfrak{z}_0 + W^l t)(x), \quad j \in \mathbb{Z}_V,$$

where $\nu_j^l = 2\Omega_{2l+1}(P_j)$ and Ψ_j is given by (6.14). Similar to the KdV-case, we find that

$$\nu_j^l = 2(i/2)^{2p+1}j^{2l+1} + O(j^{2l-3}), \quad j \in \mathbb{N}_V. \tag{6.20}$$

The system of Floquet solutions $\{v_j^l\}$ is complete. Indeed, the items of Definition 3 from 1) through 2b) describe properties of the sections Ψ_j which are the same as for the KdV equation, so we have already checked them. The property 2c) with $-\widetilde{\Delta} = -\widetilde{\Delta}^l = 2l - 3$ follows from (6.20).

The linearised lth equation satisfies the assumption v): its flow-maps $S_{\tau*}^t$ are well-defined linear isomorphisms. Indeed, by Lemma 8, outside the singular set $R_s \times \mathbb{T}^n$ the vectors $\{\Psi_j(r, \mathfrak{z})\}$ form an equivalent complex basis of the skew-normal space $T_u^{\perp c}\mathcal{T}^{2n} \subset Z_d$, where $d \geq 1$ and $u = \Phi_0(r, \mathfrak{z})$. After we choose these bases in the spaces $T_{u_0(\tau)}^{\perp c}\mathcal{T}^{2n}$ and $T_{u_0(t)}^{\perp c}\mathcal{T}^{2n}$, the map $S_{\tau*}^t$ becomes diagonal with the unit diagonal elements $\{e^{i\nu_j^l(r)(t-\tau)}\}$. So $\|S_{\tau*}^t(u_0(\tau))\|_{d,d} \leq C(r, d)$ for all τ and t, if $d \geq 1$ and $r \notin R_s$.

6.4 Linearised SG equation

Let us consider the SG equation, linearised about any finite-gap solution $u(t, x)$ as in (4.7), (4.9). Now it is more convenient to study the SG equation in the form (3.3). Accordingly, the linearised equation takes the form:

$$\ddot{\tilde{u}} = \tilde{u}_{xx} - (\cos u(t, x))\tilde{u}, \quad \tilde{w} = -A^{1/2}\dot{\tilde{u}}. \tag{6.21}$$

As in the KdV-case, solutions of equations (6.1), i.e. of the nonautonomous $\mathcal{A}$-equation and its adjoint, are given by explicit theta-formulas (see [EFM, BT]). Accordingly, Theorem 3 provides us with explicit formulas for solutions of equation (6.21) which satisfy the periodic/antiperiodic boundary conditions. Similar to the KdV-case, these solutions have the Floquet form:

$$\tilde{u}_j(t, x; D, r) = G(\tilde{V}x + Wt + D; P_j)e^{(\Omega_1 + \Omega_2)(P_j)\tilde{W} + (\Omega_1 - \Omega_2)(P_j)t},$$

$$\tilde{w}_j = -A^{-1/2}\dot{\tilde{u}}_j; \quad j \in \mathbb{Z}_n. \tag{6.22}$$

Here $G(q; p)$ is some analytic function of $q \in \mathbb{T}^n$ and $P \in \Gamma$, expressed via the θ-function, and $\Omega_1(P)$, $\Omega_2(P)$ are Abel integrals on Γ. The points P_j are defined as solutions for the following equation:

$$\frac{1}{i}(\Omega_1 + \Omega_2)(P_j) = j, \quad j \in \mathbb{Z}_V,$$

(cf. (6.12) and [EFM]). Similar to the KdV-case, the Floquet exponents $\nu_j(r) = \frac{1}{i}(\Omega_1 - \Omega_2)(P_j)$ can be described as the last components of frequency vectors $\tilde{W} \in \mathbb{R}^{n+1}$ corresponding to surfaces $\mathcal{T}^{2n+2}$, where except the previously opened gaps $[E_1, \bar{E}_1], [E_1^{-1}, \bar{E}_1^{-1}], \ldots, [E_n^{-1}, \bar{E}_n^{-1}]$ we also ε-open closed gaps at the points P_j and $1/P_j$, and next sent ε to zero.

In striking difference with the KdV case, some exponents ν_j are non-real if the open gap are sufficiently large; corresponding functions $r \mapsto \nu_j(r)$ have algebraic singularities, see [McK, EFM].

Asymptotical analysis of solutions (6.22) show that after multiplying them by proper constants the solutions take the form

$$\tilde{u}_j = \frac{1}{2\pi} e^{i(jx + \nu_j(r)t)} (1 + \zeta_j(r, D)(x)), \quad \tilde{w}_j = \ldots, \qquad (6.23)$$

where the function ζ_j goes to zero when $j \to \infty$ or $r \to 0$.

Due to the same arguments as in the KdV case, the system of Floquet solutions (6.23) satisfies the asymptotic assumption 2) from Definition 3, and due to (4.10) it is nonresonant (see [BiK1, BoK2, KK]). Using the Corollary from Lemma 10, we get that this system of Floquet solutions of the linearised equation (SG)+(OP) is complete nonresonant.

7   Normal form

7.1   A normal form theorem

We continue to study the Hamiltonian equation (5.1) near an invariant manifold $\mathcal{T}^{2n} = \Phi_0(R \times \mathbb{T}^n)$ which possesses the properties (i)–(v) as in Section 3.

Proposition 2 puts the linearised equation (5.2) to a constant coefficient normal form, provided that this equation possesses a complete non-resonant system of Floquet solutions. In this section we show that under these assumptions the equation (5.1) itself can be put to a convenient normal form in a neighbourhood of $\mathcal{T}^{2n}$. Namely, we show that the action-angle variables (p, q) on $\mathcal{T}^{2n}$ can be supplemented by a skew-orthogonal to $\mathcal{T}^{2n}$ vector-coordinate y in such a way that in the new coordinates the equation is Hamiltonian with a hamiltonian

$$h(p) + \frac{1}{2}\langle B(p)y, y \rangle + h_3(p, q, y), \quad h_3 = O(\|y\|^3).$$

Here $B(p)$ is the self-adjoint operator from Proposition 2 and the term h_3 defines a hamiltonian vector field of the same order as the nonlinear part $J\nabla H$ of the original equation (this is a crucial property of the normal form!). The normal form is an effective technical tool to study equation (5.1) and its perturbations in the vicinity of $\mathcal{T}^{2n}$.

We assume that the linearised equation (5.2) has a complete family of skew-orthogonal Floquet solutions $v_j(t)$ as in (5.4), define the singular subset R_s, $R_s = R_s^c \cap R$ as in Section 5 (see there a remark after Definition 3). As in Section 5, we choose any domain R_1 in a compact part of the regular set $R_0 = R \setminus R_s$.

By Lemma 5 the equation (3.1) is integrable in $\Phi_0(R_0 \times \mathbb{T}^n)$. So we can cover $\Phi_0(R_1 \times \mathbb{T}^n)$ by a finite system of open sub-domains such that in each one the equation admits analytic action-angle variables (p, q). For simplicity we suppose that the action-angles exist globally in $\Phi_0(R_1 \times \mathbb{T}^n)$ and use them instead of $(r, \mathfrak{z})$. Accordingly, we write $\Phi_0(R_1 \times \mathbb{T}^n)$ as $\Phi_0(P \times \mathbb{T}^n)$, where $P = \{p\} \Subset \mathbb{R}^n$ and $\mathbb{T}^n = \{q\}$.

We denote $W = P \times \mathbb{T}^n$ and $\Phi_0(W) = \mathcal{T}_1^{2n}$. The map $\Phi_0\, W \to \mathcal{T}^{2n} \subset Z$ analytically extends to a bounded analytic map $W^c \to Z^c$, where W^c is a complex neighbourhood of W, $W^c = (P + \delta) \times \{q \in \mathbb{C}^n/2\pi\mathbb{Z}^n \mid |\,\mathrm{Im}\, q| < \delta\}$.

Since $\omega = \nabla h$ (see Lemma 5), then we write the skew-orthogonal Floquet solutions $v_j(t)$ as

$$v_j(t; p, q) = e^{i\nu_j(p)t}\Psi_j(p, q + t\nabla h(p)), \quad p \in P,\ q \in \mathbb{T}^n,\ j \in \mathbb{Z}_n. \tag{7.1}$$

The linear in y map $y \mapsto \Phi_1(p, q + t\nabla h)y$ as in (5.8) reduces the linearised equation (5.2) to the constant-coefficient linear equation

$$\dot{y} = JB(p)y, \ldots, \tag{7.2}$$

where the dots stand for components of the linearised equation in directions tangent to $W \times \{0\}$ (see Proposition 2). We denote by $\mathcal{S}_\delta = \mathcal{S}_\delta(Y_d)$ the manifold

$$\mathcal{S}_\delta = W \times O_\delta(Y), \qquad Y = Y_d,$$

and denote by $\mathcal{S}_\delta^c$ its complex neighbourhood $\mathcal{S}_\delta^c = W^c \times O_\delta(Y^c)$. We give $\mathcal{S}_\delta$ symplectic structure by means of the 2-form $(dp \wedge dq) \oplus \alpha_2^Y$, where $\alpha_2^Y = \alpha_2|_Y$.

Our goal in this section is to prove the following Normal Form Theorem:

Theorem 5. *Let the Hamiltonian equation (5.1) and its invariant submanifold $\mathcal{T}^{2n}$ satisfy the assumptions i)–v); let a sub-domain $\mathcal{T}_1^{2n} = \Phi_0(P \times \mathbb{T}^n) \subset \mathcal{T}_0^{2n}$ be as above and (7.1) be a complete system of skew-orthogonal Floquet solutions of the linearised equation (5.2). Then there exists $\delta_1 > 0$ and an analytic symplectomorphism $G(\mathcal{S}_{\delta_1}, dp \wedge dq \oplus \alpha_2^Y) \to (Z, \alpha_2)$ such that $G(\mathcal{S}_{\delta_1})$ is a neighbourhood of $\mathcal{T}_1^{2n}$ and*

$$\mathcal{H} \circ G = h(p) + \tfrac{1}{2}\langle B(p)y, y \rangle + h_3(p, q, y).$$

Here $h_3 = O(\|y\|^3)$ is an analytic functional such that its gradient map is of order $\tilde{d} = \max\{d_H, -\Delta - d_J, -\tilde{\Delta} - d_J\}$, i.e. $\|\nabla_y h_3(p, q, y)\|_{s-\tilde{d}} \leq C\|y\|_s^2$ for any $(p, q, y) \in \mathcal{S}_{\delta_1}$.

To simplify our presentation we suppose below that all the frequencies $\nu_j(p)$ are real and consequently the operator $B(p)$ is diagonal in the φ_j-basis of the space Y, i.e., $B(p)\varphi_j = \frac{\nu_j(p)}{\nu_j^Y}\,\varphi_j$ for every j.

We start with the affine in $y \in Y$ map Φ,

$$\Phi = \Phi_0 + \Phi_1 \, \mathcal{S}_\delta^c \to Z^c, \qquad (p, q, y) \mapsto \Phi_0(p, q) + \Phi_1(p, q)y.$$

It is real (it sends $\mathcal{S}_\delta$ to Z), bounded on bounded subsets of $\mathcal{S}_\delta^c$ and is weakly analytic by assumptions b) and d). So Φ is an analytic map by the criterion of analyticity. By Lemma 8 its linearizations at points from $W \times \{0\}$ define isomorphisms of $\mathbb{R}^{2n} \times Y$ and Z. Thus, by the inverse function theorem the map Φ defines an analytic isomorphism of $\mathcal{S}_\delta^c$ and a complex neighbourhood of $\mathcal{T}_1^{2n}$ in Z, provided that δ is sufficiently small

Next we study symplectic properties of the map Φ. Since restriction of Φ to $W \times \{0\}$ equals Φ_0 and restriction to any disc $\{w\} \times O_\delta(Y)$ equals $\Phi_1(w)$ up to a translation, then these restrictions are symplectic. In particular, for any $w \in W$ the map $\Phi_*(w, 0)$ is a linear symplectomorphism. Hence, the pull-back form $\omega_2 := \Phi^* \alpha_2$ equals $(dp \wedge dq) \oplus \alpha_2^Y$ for $w = 0$ and these two forms coincide being restricted to any disc $\{w\} \times O_\delta(Y)$. It means that the difference $\omega_\Delta = \omega_2 - dp \wedge dq \oplus \alpha_2^Y$ may be written as

$$\omega_\Delta = j_{WW}(w, y)dw \wedge dw + j_{WY}(w, y)dy \wedge dw + j_{YW}(w, y)dw \wedge dy,$$

where the linear operators $j_{WW}(w, y)$, $j_{WY}(w, y)$ and $j_{YW}(w, y)$ vanish for $y = 0$.

In the calculations we carry out below it is convenient to adopt gradient-notations for linearizations of the maps Φ and Φ_1 in w. Namely, below we write

$$\Phi(w, y)_*(\delta w, 0) = \sum \nabla_{w_j} \Phi(w, y)\delta w_j =: \nabla_w \Phi \cdot \delta w.$$

Here $\nabla_w \Phi = (\nabla_p \Phi, \nabla_q \Phi) \in Z \times \cdots \times Z$ ($2n$ times). Similar we write $\Phi_{1*}(\delta w, 0) = \nabla_w \Phi_1 \cdot \delta w$, where any $\nabla_{w_j} \Phi_1$ is a linear operator $Y \to Z$. In these notations we have:

$$\omega_2[\delta y, \delta w] = \alpha_2[\Phi_1 \delta y, \Phi_{0*} \delta w + (\nabla_w \Phi_1 \cdot \delta w)y] = \langle \bar{J} \Phi_1 \delta y, \nabla_w \Phi_1 y \rangle \cdot \delta w$$

and $\omega_2[\delta w, \delta y] = \langle \bar{J}(\nabla_w \Phi_1 \cdot \delta w)y, \Phi_1 \delta y \rangle$. Hence,

$$\begin{aligned} j_{WY}(w, y)\delta y &= \langle \bar{J} \Phi_1(w)\delta y, \nabla_w \Phi_1(w)y \rangle, \\ j_{YW}(w, y)\delta w &= \Phi_1^*(w)\bar{J}(\nabla_w \Phi_1 \cdot \delta w)y. \end{aligned} \tag{7.3}$$

Abbreviating $(\delta w, \delta y) \in \mathbb{R}^{2n} \times Y$ to δ_3, we write the form ω_Δ as $\omega_\Delta = \bar{J}_\Delta d_3 \wedge d_3$, where $\bar{J}_\Delta$ is the operator matrix:

$$\bar{J}_\Delta = \bar{J}_\Delta(w, y) = \begin{bmatrix} j_{WW} & j_{WY} \\ -j_{WY}^* & 0 \end{bmatrix}. \tag{7.4}$$

The form ω_Δ is exact, as well as the forms α_2 and $dp \wedge dq \oplus w_2^Y$, i.e. $\omega_\Delta = d\omega_1$. Lemma 2 represents the 1-form ω_1 as

$$\begin{aligned} \omega_1(w, y) &= \left(\int_0^1 \langle \bar{J} \Phi_1(w)y, \nabla_w \Phi_1(w)ty \rangle dt \right) dw \\ &= \tfrac{1}{2} \langle \bar{J} \Phi_1(w)y, \nabla_w \Phi_1(w)y \rangle dw = \tfrac{1}{2} \alpha_2[\Phi_1(w)y, \nabla_w \Phi_1(w)y]dw. \end{aligned}$$

We have seen that

$$\omega_2 = \Phi^* \alpha_2 = (dp \wedge dq) \oplus \alpha_2^Y + d(L(w,y)dw), \qquad (7.5)$$

where the $2n$-vector L has the components $L_j \doteq \frac{1}{2}\alpha_2[\Phi_1(w)y, \nabla_{w_j}\Phi_1(w)y]$.

Next we calculate how the map Φ changes the hamiltonian $\mathcal{H}$. To begin with we analyse how Φ_1 transforms the quadratic part $\langle Au, u \rangle$ of the hamiltonian $\mathcal{H}$.

Since the nonautonomous symplectic linear map $\Phi^t := \Phi_1(p, q + t\nabla h)$ sends solutions $y(t)$ of equation (7.2) to solutions $v(t) = \Phi^t y(t)$ of (5.2), then we have the equalities:

$$
\begin{array}{ccc}
\dot{v} & = & \dot{\Phi}^t y + \Phi^t \dot{y} \\
\| & & \| \\
J A_t v & & \dot{\Phi}^t y + \Phi^t J B(p) y \\
\| & & \\
J A_t \Phi^t y & &
\end{array}
$$

Thus,

$$J A_t \Phi^t y = \dot{\Phi}^t y + \Phi^t J B(p) y. \qquad (7.6)$$

Taking skew-product of (7.6) with $-v$, we get:

$$
\begin{array}{ccc}
\langle J A_t \Phi^t y, \bar{J} v \rangle & = & \langle \dot{\Phi}^t y + \Phi^t J B(p) y, \bar{J} v \rangle \\
\| & & \| \\
\langle \Phi^{t*} A_t \Phi^t y, y \rangle & & \langle \dot{\Phi}^t y, \bar{J}\Phi^t y \rangle + \langle B(p)y, y \rangle,
\end{array}
\qquad (7.7)
$$

where we use that $\langle \Phi^t J B y, \bar{J}\Phi^t y \rangle = \langle J B y, \bar{J} y \rangle = \langle B y, y \rangle$ by symplecticity of the map Φ^t.

Since for $t = 0$ we have $A_t = A + \nabla H(\Phi_0(w))_*$ and $\dot{\Phi}^t = \nabla_q \Phi_1(w) \cdot \nabla h(p)$, then relation (7.6) with $t = 0$ implies that

$$\Phi_1(w) J B(p) = J(A + \nabla H(\Phi_0(w))_*)\Phi_1(w) - \nabla_q \Phi_1(w) \cdot \nabla h(p).$$

Similar, (7.7) implies that

$$\langle (B(p) - \Phi_1(w)^*(A + \nabla H(\Phi_0(w))_*)\Phi_1(w))y, y \rangle$$
$$= \langle \Phi_1(w)^* \bar{J}(\nabla_q \Phi_1(w) \cdot \nabla h(p))y, y \rangle = \langle \mathfrak{A}(w)y, y \rangle,$$

where $\mathfrak{A}$ stands for the symmetrisation of the operator $\Phi_1^* \bar{J}(\nabla_q \Phi_1 \cdot \nabla h)$, i.e.,

$$\mathfrak{A}(w) = \frac{1}{2}\left(\Phi_1^*(w)\bar{J}(\nabla_q \Phi_1(w) \cdot \nabla h(p)) - (\nabla_q \Phi_1(w)^* \cdot \nabla h(p))\bar{J}\Phi_1(w) \right).$$

Since this relation holds identically in $y \in Y$, then

$$(B(p) - \Phi_1(w)^*(A + \nabla H(\Phi_0(w))_*)\Phi_1(w) = \mathfrak{A}(w).$$

Lemma 11. *The operator $\mathfrak{A}$ defines a $(\Delta + d_J)$-smoothing symmetric map $\mathfrak{A} : Y_d^c \to Y_{d+d_J+\Delta}^c$, analytic in $w \in W^c$.*

Proof. The operator $\mathfrak{A}$ is symmetric by its construction. It remains to check its smoothness.

Since $\nabla_q \Phi_1 = \nabla_q(\Phi_1 - \iota)$, then by (5.10) and the Cauchy estimate the operator $\nabla_q \Phi_1 \cdot \nabla h$ analytically depends on $w \in W^c$ as a map $Y_d^c \to Z_{d+\Delta}$. By Lemma 9 the operator $\Phi_1(w)^* \bar{J} : Z_{d+\Delta}^c \to Y_{d+d_J+\Delta}^c$ also is analytic in w. Hence, the first term of the operator $\mathfrak{A}$ defines an analytic in $w \in W^c$ map $Y_d^c \to Y_{d+d_J+\Delta}^c$.

Using Lemma 9 once again we find that the operator $\bar{J}\Phi_1(w) : Y_d^c \to Z_{d+d_J}^c$ is analytic in w. Due to the second assertion of this lemma and the Cauchy estimate, the map $\nabla_q \Phi_1^* \cdot h : Z_{d+d_J}^c \to Y_{d+d_J+\Delta}^c$ is analytic in w as well. Combining these two statements, we find that the second term of $\mathfrak{A}$ also defines an analytic in $w \in W^c$ map $Y_d^c \to Y_{d+d_J+\Delta}^c$. This completes the proof. $\square$

Now we write the transformed hamiltonian as $\mathcal{H} \circ \Phi = \frac{1}{2}\langle A\Phi_0, \Phi_0\rangle + \langle A\Phi_0, \Phi_1 y\rangle + \frac{1}{2}\langle A\Phi_1 y, \Phi_1 y\rangle + H(\Phi)$, and separate its affine in y part:

$$\mathcal{H} \circ \Phi = \left(\frac{1}{2}\langle A\Phi_0, \Phi_0\rangle + H(\Phi_0)\right) + \left(\langle A\Phi_0, \Phi_1 y\rangle + \langle \nabla H(\Phi_0), \Phi_1 y\rangle\right)$$
$$+ \left(\frac{1}{2}\langle A\Phi_1 y, \Phi_1 y\rangle + H(\Phi_0 + \Phi_1 y) - H(\Phi_0) - \langle \nabla H(\Phi_0), \Phi_1 y\rangle\right).$$

The first term in the right-hand side equals $h(p)$.

By Lemma 2 the form $\omega_2 = \Phi^* \alpha_2$ equals $(dp \wedge dq) \oplus \alpha_2^Y$, when $y = 0$. Hence, for $y = 0$ the y-component of equation (5.1), written in the (p, q, y)-variables, is $J\nabla_y(\mathcal{H} \circ \Phi)$. It equals zero since the set $\{y = 0\}$ is invariant for the equation. Thus, the second term vanishes.

By Lemma 11, $\langle A\Phi_1 y, \Phi_1 y\rangle = \langle By, y\rangle - \langle \nabla H_* \Phi_1 y, \Phi_1 y\rangle - \langle \mathfrak{A}y, y\rangle$. Therefore the third term in the r.h.s. equals $\frac{1}{2}\langle B(p)y, y\rangle + h_2(p, q, y)$, where

$$h_2 = -\frac{1}{2}\langle \nabla H(\Phi_0)_* \Phi_1 y, \Phi_1 y\rangle - \frac{1}{2}\langle \mathfrak{A}(w)y, y\rangle$$
$$+ H(\Phi_0 + \Phi_1 y) - H(\Phi_0) - \langle \nabla H(\Phi_0), \Phi_1 y\rangle.$$

It is easy to see, using Lemmas 9 and 11, that h_2 defines an analytic gradient map $\nabla_y h_2 \, \mathbb{R}^{2n} \times Y_d \to Y_{d-\bar{d}}$.

Thus, the affine in y map Φ transforms the hamiltonian $\mathcal{H}$ to

$$\mathcal{H} \circ \Phi = h(p) + \frac{1}{2}\langle B(p)y, y\rangle + h_2(p, q, y),$$

where $h_2 = O\|y\|^2$ and $\operatorname{ord} \nabla h_2 = \tilde{d}$.

Our next goal is to normalise the symplectic structure $\omega_2 = \Phi^* \alpha_2$ in $\mathcal{S}_\delta$ by means of the Moser–Weinstein theorem (Lemma 3). The theorem states that

$\varphi^*\omega_2 = (dp \wedge dq) \oplus \alpha_2^Y$, where φ is the time-one shift S_0^1 along trajectories of a nonautonomous equation:

$$\dot{\mathfrak{z}} = V^t(\mathfrak{z}), \quad \mathfrak{z} = (w, y).$$

The vector field $V^t \, \mathcal{S}_\delta \to \mathbb{R}^{2n} \times Y_d$ is obtained as a solution of the equation

$$-(\overline{J}_0 + t\overline{J}_\Delta)V^i = a(\mathfrak{z}, y),$$

where $\overline{J}_0(\delta p, \delta q, \delta y) = (-\delta q, \delta p, \overline{J}\delta y)$, the operator $\overline{J}_\Delta$ is as in (7.4) and the map a is such that differential of the 1-form $a(\mathfrak{z})d\mathfrak{z}$ equals $\omega_2 - (dp \wedge dq) \oplus \alpha_2^Y$. By (7.5), $a(\mathfrak{z}) = (L(\mathfrak{z}), 0)$.

We claim that the map φ sends $\mathcal{S}_{\delta_1}^c$ to $\mathcal{S}_\delta^c$ (δ_1 is sufficiently small compare to δ) and transforms $\mathcal{H} \circ \Phi$ to a hamiltonian of similar form:

Lemma 12. *The hamiltonian $\mathcal{H} \circ \Phi \circ \varphi$ equals to*

$$\mathcal{H} \circ \Phi \circ \varphi = h(p) + \tfrac{1}{2}\langle B(p)y, y \rangle + \tfrac{1}{2}\langle \mathfrak{B}(p, q)y, y \rangle + h_3(p, q, y), \qquad (7.8)$$

where $B(p)$ is the same as in (7.2) and $\mathfrak{B}(p, q)$ is a linear operator of order $\tilde{d}$, analytic in (p, q) ($\tilde{d}$ is the same as in Theorem 5)). The function $h_3 = O(\|y\|^3)$ has an analytic gradient map of order $\tilde{d}$, $\|\nabla_y h_3(p, q, y)\|_{d+\tilde{d}} \leq C\|y\|^2$.

The statement of the lemma is quite obvious for a finite-dimensional phase space $\mathcal{S}_\delta$, but not in the infinite-dimensional situation. Indeed, the transformation φ has the form $\varphi = \mathrm{id} + \widetilde{\varphi}$, where $\widetilde{\varphi} = \|y\|^2$ is a Δ-smoothing map. Thus the transformed hamiltonian gets the term $\langle B(p)y, \widetilde{\varphi}y \rangle$ which is $O(\|y\|^3)$ with the gradient of order $d_A - \Delta$. The number $d_A - \Delta$ could be rather big and the term could spoil the forthcoming constructions. Fortunately, it vanishes up to a smoother term. This is essentially what the lemma states. For its proof see Lemma 7 in [K1] and [KK].

To prove the theorem it remains to check that the operator $\mathfrak{B}$ in (7.8) vanishes. Since φ is analytic and $O(\|y\|^2)$-close to the identity, then $\varphi_*(w, 0)|_{\{0\} \times Y} = \mathrm{id}$. Thus the transformation $y(t) \mapsto (\Phi \circ \varphi)_*(p, q + t\nabla h, 0)$ $y(t)$ sends solutions of the equation (7.2) to solutions of (5.2).

From other side, $\Phi \circ \varphi$ is a canonical transformation which transforms solutions of the equation with hamiltonian (7.8) to solutions of (5.1). In particular, it sends the curves $w(t; p, q) = (p, q + t\nabla h(p), 0)$ to solutions $u_0(t)$ of (5.1). Hence, the linearisation $(\Phi \circ \varphi(w(t)))_*$ transforms solutions of the linearised equation

$$\dot{y} = J(B(p) + \mathfrak{B}(w(t))y, \dots \qquad (7.9)$$

to solutions of (5.2). Therefore the transformation $\varphi(w(t))_*$ sends solutions of (7.9) to solutions of (7.2). Since a y-component of the map $\varphi(w(t))_*$ is the identity, then we must have $JB(p)y \equiv J(B(p) + \mathfrak{B}(p, q + t\nabla h))y$. This implies that $\mathfrak{B} \equiv 0$ and the theorem is proven. $\square$

7.2   Examples

1) *Korteweg–de Vries equation.* The KdV equation in a Sobolev space $Z_s = H_0^s(S^1)$ with $s \geq 3$, which is given symplectic structure by the form $\alpha_2 = \langle (-\partial/\partial x)^{-1} du, du \rangle_{L_2}$ takes the form (3.1) (see Example 3.1). Its restriction to a bounded part $\mathcal{T}^{2n}$ of any finite-gap manifold $\mathcal{T}_V^{2n}$ (see (4.5)) satisfies the restrictions i)–v) and the corresponding system of Floquet solutions (6.13) is complete non-resonant with $\Delta = \widetilde{\Delta} = 1$, $d_J = 1$, $d_H = 0$ and $d_A = 2$. Therefore Theorem 5 provides KdV with a normal form. To state the result, we find the singular subset $R_s \subset R$ [15] and choose any domain $R_1 \Subset R \setminus R_s$. We cover R_1 up to its zero-measure subset by non-overlapping sub-domains $R_{11}, R_{12}, \ldots$ such that the KdV-equation restricted to any manifold $\Phi_0(R_{1j} \times \mathbb{T}^n) = J_j^{2n}$ admits action-angle variables (p, q) with $p \in P_j \Subset \mathbb{R}^n$.

For any s we denote by $Y_s \subset H_0^s(S^1)$ the closed subspace spanned by the functions $\{\cos jx, -\sin jx \mid j \in \mathbb{N}_V\}$. Applying Theorem 5 we get:

Theorem 11. *For any $s \geq 3$ there exists $\delta > 0$ and an analytic symplecto-morphism G of $(P_j \times \mathbb{T}^n \times \{\|y\|_s < \delta\}, dp \wedge dq \oplus \alpha_2^Y)$ and a domain in (H_0^s, α_2). It contains $\mathcal{T}_j^{2n}$ in its range and G^{-1} transforms KdV to the Hamiltonian system*

$$\dot{q} = \nabla_p \mathcal{H}, \quad \dot{p} = -\nabla_q \mathcal{H}, \quad \dot{y} = \frac{\partial}{\partial x} \nabla_y \mathcal{H} \tag{7.10}$$

with a hamiltonian $\mathcal{H}$ of the form $\mathcal{H} = h(p) + \frac{1}{2}\langle B(p)y, y \rangle + H_3(p, q, y)$. Here $h(p)$ is the hamiltonian of KdV restricted to $\mathcal{T}_j^{2n}$, $B(p)$ is the linear operator in Y_d with eigenvectors $\cos mx, -\sin mx$ and eigenvalues $\nu_m(p)$ $(m \in \mathbb{N}_V)$, $H_3 = O(\|y\|_s^3)$ is a function with a zero-order analytic gradient map.

2) *Higher KdV equations.* Let us take any lth equation from the KdV-hierarchy. Since the same (as in the KdV-case) sections Ψ_j of the skew-normal bundle to a finite-gap manifold $\mathcal{T}_V^{2n}$ give rise to its Floquet solutions, then the same map Φ_1 reduces the linearised lth equation to the equation $\dot{y} = JB^l(p)y$ in the space Y. Here $J = \partial/\partial x$ and $B^l(p)$ is a linear operator with the eigenvectors $\cos jx, -\sin jx$, corresponding to the eigenvalues $\nu_j^l(p)$ $(j \in \mathbb{N}_V)$ as in (6.20) Therefore, the same map G with $s \geq 2p + 1$ reduces the lth KdV equation in the vicinity of $\mathcal{T}_j^{2n}$ (the same as above part of $\mathcal{T}_V^{2n}$) to the equation (7.10) with $\mathcal{H} = \mathcal{H}^l(p, q, y) = h_l(p) + \frac{1}{2}\langle B^l(p)y, y \rangle + H_3^l(p, q, y)$. Here h_l is the hamiltonian of the lth equation restricted to $\mathcal{T}_j^{2n}$ and the operator $B^l(p)$ has the eigenvalues ν_j^l. Now $\Delta = d_J = 1$ as in the KdV-case, $d_A = 2l$, $d_H = 2l - 2$ and $\widetilde{\Delta} = 3 - 2l$ by (6.20). So $\bar{d} = 2l - 2$ and $H_3 = O(\|y\|_s^3)$ has an analytic gradient map of order $2l - 2$.

3) *Sine-Gordon equation.*

---

[15] In the KdV-case the set R_s is empty. We neglect this nice specificity of KdV since our goal is to present arguments applicable to other integrable PDEs.

8   The KAM theorem

8.1   The main theorem and related results

Let $(\{Z_s\}, \alpha_2)$, $\alpha_2 = \overline{J} dz \wedge dz$ be a scale of symplectic Hilbert spaces and ord $\overline{J} = -d_J \leq 0$. Let $\mathcal{H}$ be a quasilinear hamiltonian of the form $\mathcal{H} = \frac{1}{2}\langle Az, z \rangle + H(z)$, where A is a selfadjoint isomorphism of the scale of order $d_A > -d_J$. We fix any $d \geq d_A/2$ and assume that the function H is analytic in the space Z_d (or in a neighbourhood in Z_d of the manifold $\mathcal{T}_0^{2n}$, see below) and defines an analytic gradient map of order d_H, $\nabla H : Z_d \to Z_{d-d_H}$. We have $d_H < d_A$ due to the quasilinearity of the hamiltonian $\mathcal{H}$. The corresponding Hamiltonian equation takes the form:

$$\dot{u} = J\nabla\mathcal{H}(u) = J(Au + \nabla H(u)), \quad J = (-\overline{J})^{-1}. \tag{8.1}$$

As in Sections 3 and 5 we assume that the equation (8.1) has an invariant manifold $\mathcal{T}_0^{2n} = \Phi_0(R_0 \times \mathbb{T}^n)$ filled with quasiperiodic solutions $u_0(t; r, \mathfrak{z})$ which satisfies the assumptions i)–v). The manifold R_0 is the regular part of an n-dimensional real analytic set R. By $\widetilde{R}$ we denote any chart on R_0 analytically diffeomorphic to a bounded connected subdomain of $\mathbb{R}^n$. We identify $\widetilde{R}$ with this domain and supply it with the n-dimensional Lebesgue measure mes_n.

As in Section 5, we also consider linearisation of the equation (8.1) about a solution u_0:

$$\dot{v} = J\big(Av + \nabla H(u_0(t))_* v\big), \tag{8.2}$$

and assume that (8.2) has a system (5.4) of Floquet solutions $v_j(t)$.

Our concern in this section is a hamiltonian perturbation of the equation (8.1):

$$\dot{u} = J(Au + \nabla H(u) + \varepsilon \nabla H_1(u)), \tag{8.3}$$

and behaviour of solution for (8.3) near the manifold $\mathcal{T}_0^{2n}$. We assume that H_1 is an analytic functional such that ord $\nabla H_1 = d_H$.

By $\tilde{d}$ we denote the real number from Theorem 5, $\tilde{d} = \max\{d_H, -\Delta - d_J, -\widetilde{\Delta} - d_J\}$, where $-\Delta$ is the order of the linear operator $\Phi_1 - \iota$ (see (5.10)) and $\widetilde{\Delta}$ is the exponent of growth in j of "variable parts" of the the Floquet exponents $\nu_j(r)$ (see (5.12)).

Let us fix any $\tilde{\rho}$ such that $0 < \tilde{\rho} < 1/3$. Now we state a KAM theorem which is the main result of this paper:

Theorem 7 (the Main Theorem). *Let the invariant manifold $\mathcal{T}_0^{2n}$ satisfy the assumptions i)–v) and the system of Floquet solutions (5.4) for (8.3) is complete nonresonant. Besides, $d_1 := d_A + d_J \geq 1$ and*

1) (spectral asymptotic): $\nu_j(r) = K_1 j^{d_1} + K_1^1 j^{d_1^1} + K_1^2 j^{d_1^2} + \cdots + \tilde{\nu}_j(r)$, where $K_1 > 0$, $d_1 > d_1^1 > \ldots$ (the dots stand for a finite sum), the functions $\tilde{\nu}_j$ analytically extend to R^c, where $|\tilde{\nu}_j| \leq Cj^\kappa$ with some $\kappa < d_1 - 1$;

2) (quasilinearity): $\tilde{d} < d_1 - 1$.

Then most of the invariant tori $T^n(r)$ of equation (8.1) persist in (8.3) when $\varepsilon \to 0$ in the following sense: for any chart $\tilde{R} \subset R_0$ as above and any sufficiently small $\varepsilon > 0$, a Borel subset $\tilde{R}_\varepsilon \Subset \tilde{R}$ and a Lipschitz embedding $\Sigma^\varepsilon : \tilde{R}_\varepsilon \times \mathbb{T}^n \longrightarrow Z_d$ can be found such that:

a) $mes_n(\tilde{R} \setminus \tilde{R}_\varepsilon) \longrightarrow 0$ as $\varepsilon \longrightarrow 0$,

b) the map $(\Sigma^\varepsilon - \Phi_0) : \tilde{R}_\varepsilon \times \mathbb{T}^n \longrightarrow Z_d$ is bounded by $C\varepsilon^{\tilde{\rho}}$ as well as its Lipschitz constant and is analytic in $q \in \mathbb{T}^n$;

c) each torus $T_\varepsilon^n(r) := \Sigma^\varepsilon(\{r\} \times \mathbb{T}^n)$, $r \in \tilde{R}_\varepsilon$, is invariant for the equation (8.3) and is filled with its time-quasiperiodic solutions $\mathfrak{h}_\varepsilon(t)$ of the form $\mathfrak{h}_\varepsilon(t) = \mathfrak{h}_\varepsilon(t; r, \mathfrak{z}) = \Sigma^\varepsilon(r, \mathfrak{z} + t\omega_\varepsilon(r))$, where $|\omega_\varepsilon - \omega| + Lip(\omega_\varepsilon - \omega) \leq C\varepsilon^{\tilde{\rho}}$.

Amplification. *The statements b), c) of Theorem 7 remain true with $\tilde{\rho}$ replaced by any $\rho' < 1$. Besides, $\|\Sigma^\varepsilon - \Phi_0\|_d + |\omega_\varepsilon - \omega| \leq C\varepsilon$.*

We denote $\tilde{\mathcal{T}}^{2n} = \Phi_0(\tilde{W})$, $\tilde{W} = \tilde{R} \times \mathbb{T}^n$ and $\tilde{\mathcal{T}}_\varepsilon^{2n} = \Sigma^\varepsilon(\tilde{W}_\varepsilon)$, $\tilde{W}_\varepsilon = \tilde{R}_\varepsilon \times \mathbb{T}^n$. The set $\tilde{\mathcal{T}}_\varepsilon^{2n}$ is a remnant of the invariant manifold $\tilde{\mathcal{T}}^{2n}$ in the perturbed equation (8.3).

Since $\tilde{\mathcal{T}}^{2n}$ is a $2n$-dimensional manifold embedded to Z_d, then its $2n$-dimensional Hausdorff measure $mes_{2n}^{\mathcal{H}} \tilde{\mathcal{T}}^{2n}$ is finite and positive. The remnant set $\tilde{\mathcal{T}}_\varepsilon^{2n}$ is very irregular (it is totally disconnected). Still it carries most of a measure of the set $\tilde{\mathcal{T}}^{2n}$:

Proposition 4. *Under the assumptions of Theorem 7, $mes_{2n}^{\mathcal{H}} \tilde{\mathcal{T}}_\varepsilon^{2n}$ is bigger than $mes_{2n}^{\mathcal{H}} \tilde{\mathcal{T}}^{2n} - o(1)$ as $\varepsilon \to 0$.*

Proof. [16] By the assertion a) of the theorem,

$$mes_{2n}^{\mathcal{H}}(\tilde{W} \setminus \tilde{W}_\varepsilon) = o(1). \tag{8.4}$$

The map $\Phi_0 : \tilde{W}_\varepsilon \xrightarrow{\;\sim\;} \Phi_0(\tilde{W}_\varepsilon) \subset \tilde{\mathcal{T}}^{2n}$ is Lipschitz and has a Lipschitz inverse, so the map $\Sigma^\varepsilon \circ \Phi_0^{-1} : \Phi_0(\tilde{W}_\varepsilon) \xrightarrow{\;\sim\;} \tilde{\mathcal{T}}_\varepsilon^{2n}$ has the form $id + L$, where $Lip\, L \leq C\varepsilon^{\tilde{\rho}}$ (we use the assertion b)). Therefore, $mes_{2n}^{\mathcal{H}} \tilde{\mathcal{T}}_\varepsilon^{2n} \geq mes_{2n}^{\mathcal{H}} \Phi_0(\tilde{W}_\varepsilon) - O(\varepsilon^{\tilde{\rho}})$. Since $mes_{2n}^{\mathcal{H}}(\Phi_0(W \setminus \tilde{W}_\varepsilon)) = o(1)$ by (8.4), then the assertion follows. $\square$

Under the assumptions of Theorem 7, a solution $u_0(t; r, \mathfrak{z})$ of (8.1) is linearly stable if all Floquet exponents $\nu_j(r)$ are real (see the Corollary to Proposition 2). Let us assume that this is the case for all $r \in \tilde{R}$. Then the solutions $\mathfrak{h}_\varepsilon(t; r, \mathfrak{z})$ of the perturbed equation (8.3) with $r \in \tilde{R}_\varepsilon$ also are linearly stable, provided that this equation linearised about $\mathfrak{h}_\varepsilon$ satisfies some a priori estimate. We recall that by the assumption v) the flow maps $S_{r*}^t(\mathfrak{h}_\varepsilon(\tau))$ of

---

[16] For basic properties of the Hausdorff measure which we use below see e.g. [Fal].

the linearised equation are well defined in the space Z_d. We say that the linearised equation is *uniformly well defined* (in Z_d) if

$$\|S_{\tau *}^t(\mathfrak{h}_\varepsilon(\tau))\|_{d,d} \le C_1 e^{C_2 t} \quad \text{for all } t, \tau. \tag{8.5}$$

Theorem 8. *If under the assumptions of Theorem 7 all the Floquet exponents $\nu_j(r)$ are real for all $r \in \tilde{R}$, then a solution $\mathfrak{h}_\varepsilon(t)$ is linearly stable, provided that equation (8.3) linearised about this solution is uniformly well defined.*

(Examples we consider below in Section 9 show that the assumption (8.5) is quite non-restrictive).

We prove the two theorems and the amplification, reducing them to similar statements concerning perturbations of parameter-depending linear systems.

8.2   Reduction to a parameter-depending case

We perform the reduction in four steps.

Step 1 (localisation). Let us denote by R_f the set of singularities of the frequency map ω, $R_f = \{r \in \tilde{R} \mid \det \omega_*(r) = 0\}$, and denote $\tilde{R}_s = (R_s \cap \tilde{R}) \cup R_f$, where R_s is the singular set, constructed in Section 5. By the assumption iv), R_f is a proper analytic subset of $\tilde{R}$. So $\tilde{R}_s$ also is one, and for any given positive γ_0 we can find a finite system of M connected subdomains $\tilde{R}_l \subset \tilde{R} \setminus \tilde{R}_s$ such that $\text{dist}\,(r_j, r_{j'}) \ge C(\gamma_0) > 0$ if $r_j \in \tilde{R}_j$ and $r_{j'} \in \tilde{R}_{j'}$ with $j \ne j$. Besides,

   a) $\text{mes}\,(\tilde{R} \setminus \cup \tilde{R}_l) < \gamma_0$,

   b) the hamiltonian system restricted to $\Phi_0(\tilde{R}_l \times \mathbb{T}^n)$ admits analytic action-angle variables (p, q), where $p \in P_l \Subset \mathbb{R}^n$ and $q \in \mathbb{T}^n$. The map $(p, q) \mapsto (r, \mathfrak{z})$ has the form $r = r(p)$, $\mathfrak{z} = q + \mathfrak{z}_0(p)$. This map, its inverse and the hamiltonian $h = h_l(p)$ all are δ-analytic with some positive $\delta = \delta(\gamma_0)$. By Lemma 2.2, $\nabla h(p) \equiv \omega(r(p))$;

   c) for every l the gradient map $p \mapsto \nabla h(p) \equiv \omega(r(p))$ defines a diffeomorphism $P_l \underset{\sim}{\longrightarrow} \Omega_l \Subset \mathbb{R}^n$ which is δ-analytic as well as its inverse;

   d) since each domain $\tilde{R}_l$ is connected, then the eigen-vectors ψ_j of the operator $JB(r)$ are r-independent when $r \in \tilde{R}_l$.

Step 2 (a normal form theorem). In this step of the proof and in the next Step 3 we consider any fixed domain $\tilde{R}_l$ as above and drop the index l.

Applying Theorem 5 we find an analytic symplectomorphism G which transforms the equation (8.1) in the vicinity of $\Phi_0(\tilde{R} \times \mathbb{T}^n)$ to the form given in the theorem. The same symplectomorphism converts the perturbed equation (8.3) to the Hamiltonian system

$$\dot{p} = -\nabla_q \mathcal{H}_\varepsilon, \quad \dot{q} = \nabla_p \mathcal{H}_\varepsilon, \quad \dot{y} = J\nabla_y \mathcal{H}_\varepsilon. \tag{8.6}$$

Here $p \in P$, $q \in \mathbb{T}^n$, $y \in O_\delta(Y_d)$ and $\mathcal{H}_\varepsilon = h(p) + \frac{1}{2}\langle B(p)y, y\rangle + h_3(p, q, y) + \varepsilon H_1(p, q, y)$ with $h_3 = O(\|y\|_d^3)$ and $\text{ord}\,\nabla_y h_3 = d$. The operator $B(p)$, the functions h, h_3, H_1 and their gradients all are δ-analytic in the corresponding domains.

Step 3 (introducing a parameter). Let us consider the following neighbourhoods of the torus $T_0^n = \{0\} \times \mathbb{T}^n \times \{0\}$ in $\mathcal{Y}$ and $\mathcal{Y}^c$:

$$Q_\delta = O_\delta(\mathbb{R}^n) \times \mathbb{T}^n \times O_\delta(Y_d) \subset \mathcal{Y},$$
$$Q_\delta^c = O_\delta(\mathbb{C}^n) \times \{|\mathrm{Im}\, q| < \delta\} \times O_\delta(Y_d^c) \subset \mathcal{Y}^c.$$

In the equation (8.6) we perform a shift of the action p:

$$(p, q, y) = (\tilde{p} + a,\ \tilde{q},\ \tilde{y}) =:\ \mathrm{Shift}_a(\tilde{p},\ \tilde{q},\ \tilde{y}),$$

where $a \in P$ is a parameter of the shift. After this transformation hamiltonian $\mathcal{H}_\varepsilon$ becomes an analytic function $\mathcal{H}_\varepsilon(\tilde{p},\ \tilde{q},\ \tilde{y};\ a)$ of the tilde-variables from the domain Q_δ^c. It has the following form:

$$\mathcal{H}_\varepsilon = h(a) + \nabla h(a) \cdot \tilde{p} + \tfrac{1}{2}\langle B(a)\tilde{y}, \tilde{y}\rangle + \varepsilon H_1(\tilde{p} + a,\ \tilde{q},\ \tilde{y}) + \tilde{h}_3(\tilde{p},\ \tilde{q},\ \tilde{y};\ a),$$

where $\tilde{h}_3 = O(\|\tilde{y}\|_d^3 + |\tilde{p}|^2 + |\tilde{p}|\|\tilde{y}\|_d^2)$ and $\|\nabla_y \tilde{h}_3\|_{d-\tilde{d}} = O(\|\tilde{y}\|_d^2 + |\tilde{p}|\|\tilde{y}\|_d)$ (so ord $\nabla \tilde{h}_3 = \tilde{d}$).

The functions h, H_1, $\tilde{h}_3$ and the Floquet exponents ν_j are analytic bounded functions of the parameter $a \in P + \delta$. Because the property c) from Step 1, the map $P \ni a \mapsto \omega = \nabla h(a) \in \Omega$ defines an analytic Lipschitz diffeomorphism of P and a bounded domain $\Omega \subset \mathbb{R}$. We drop the tildes and change the parameter a to ω. Now the hamiltonian $\mathcal{H}_\varepsilon$ reeds as

$$\mathcal{H}_\varepsilon(p, q, y; \omega) = h(a) + \omega \cdot p + \tfrac{1}{2}\langle B(\omega)y, y\rangle + \varepsilon H_1(p, q, y; \omega) + h_3(p, q, y; \omega).$$

The operator JB is diagonal in the symplectic basis $\{\psi_j\}$, constructed in Proposition 2:

$$J\psi_j = \pm i\nu_j^J \psi_j, \quad JB\psi_j = \nu_j(r)\psi_j \quad \forall j \in \mathbb{Z}_n.$$

Since the hamiltonian $\mathcal{H}_\varepsilon$ is δ-analytic, then by the Cauchy estimate it is Lipschitz in $a \in P$ as well as in $\omega \in \Omega$. This is all we need from its dependence in the parameters.

In the vicinity of the torus $T_0^n = \{0\} \times \mathbb{T}^n \times \{0\}$ in Q_δ the hamiltonian $\mathcal{H}_\varepsilon$ is a perturbation of the q-independent hamiltonian $\mathcal{H}_0 = \omega \cdot p + \tfrac{1}{2}\langle B(\omega)y, y\rangle$ (modulo the irrelevant constant $h(a)$). Indeed, ε is small and the term h_3 has on T_0^n a high-order zero.

The hamiltonian equations with the hamiltonian $\mathcal{H}_\varepsilon(p, q, y; \omega)$ take the form:

$$\dot{p} = -\nabla_q(\varepsilon H_1 + h_3), \quad \dot{q} = \omega + \nabla_p(\varepsilon H_1 + h_3),$$
$$\dot{y} = J(B(\omega)y + \varepsilon \nabla_y H_1 + \nabla_y h_3). \tag{8.7}$$

We abbreviate (p, q, y) to $\mathfrak{h}$ and rewrite (8.7) as $\dot{\mathfrak{h}} = V_{\mathcal{H}_\varepsilon}(\mathfrak{h})$. In the context of equations (8.7), we call the functions $\nu_j(\omega)$ *frequencies* of the linear equation.

Hamiltonian vector fields with hamiltonians of the form $\mathcal{H}_\varepsilon$ are studied in [K, K2, P]. Now we break the proof of Theorem 7 to present the main result of these works. After this we make the last step to complete the proof of Theorem 7.

8.3   A KAM-theorem for parameter-depending equations

To state the theorem we need, we relax restrictions on the hamiltonian $\mathcal{H}_\varepsilon$ as in the assumptions 1)–3) below:

1) (frequencies). The complex functions $\nu_j(\omega)$, $j \in \mathbb{Z}_n$, are Lipschitz, are real for $|j| \geq j_1$ with some $j_1 \geq n + 1$ and are odd in j, $\nu_j \equiv -\nu_{-j}$. For $j \geq n + 1$ and for some fixed $\omega_0 \in \Omega$ the following estimates hold:

$$|\nu_j(\omega_0) - K_1 j^{d_1} - K_1^1 j^{d_1^1} - K_1^2 j^{d_1^2} - \dots | \leq K j^{\tilde{d}} \quad \text{and} \quad \mathrm{Lip}\, \nu_j \leq K j^{\tilde{d}},$$

where $K_1 > 0$, $d_1 \geq 1$, $\tilde{d} < d_1 - 1$ and the dots stand for a finite sum with some exponents $d_1 > d_1^1 > d_1^2 > \dots$.

2) (perturbation). The functions h_3 and H_1 are analytic in $(p, q, y) \in Q_\delta^c$ and everywhere in Q_δ^c satisfy the estimates:

$$\left.\begin{array}{l}
|H_1| + \|\nabla_y H_1\|_{d - \tilde{d} + d_J} \leq 1 \quad \forall \omega, \\[4pt]
|h_3| \leq K(|p|^2 + |p|\|y\|_d^2 + \|y\|_d^3) \quad \forall \omega, \\[4pt]
\|\nabla_y h_3\|_{d - \tilde{d} + d_J} \leq K(|p|\|y\|_d + \|y\|_d^2) \quad \forall \omega, \\[4pt]
\text{the same estimates hold for Lipschitz constants} \\[4pt]
\text{in } \omega \in \Omega \text{ of these functions and their gradients.}
\end{array}\right\} \quad (8.8)$$

3) (domain of parameters). Ω is a bounded Borel set in $\mathbb{R}^n$ of positive Lebesgue measure, such that $\mathrm{diam}\, \Omega \leq K_2$ and $|\omega| \leq K$ for every $\omega \in \Omega$.

Let us choose any $\rho \in (0, \frac{1}{3})$.

Theorem 9. *Suppose that the assumption 1)-3) hold. Suppose also that there exist integer $j_2 \geq n$ and M_1, depending only on n, d_1, $\tilde{d}$, K, K_1, K_2 and $K_1^1, K_1^2 \dots$, such that*

$$|s \cdot \omega + l_{n+1}\nu_{n+1}(\omega) + \dots + l_{j_2}\nu_{j_2}(\omega)| \geq K_3 > 0 \qquad (8.9)$$

for all $\omega \in \Omega$, all integer n-vectors s and all j_2-vectors l such that $|s| \leq M_1$ and $1 \leq |l| \leq 2$.

Then for arbitrary $\gamma > 0$ and for sufficiently small $\varepsilon \leq \bar{\varepsilon}(\gamma)$ $(\bar{\varepsilon} > 0)$, a Borel subset $\Omega_\varepsilon \subset \Omega$ and a Lipschitz embedding $\Sigma_\varepsilon : \mathbb{T}^n \times \Omega_\varepsilon \longrightarrow Q_\delta$ can be found with the following properties:

a) mes $(\Omega \setminus \Omega_\varepsilon) \leq \gamma$,

b) the map Σ_ε is $C\varepsilon^\rho$-close to the map $\Sigma_0 : (q,\omega) \mapsto (0,q,0) \in Q_\delta$ and Lipschitz constant of the difference-map $\Sigma_\varepsilon - \Sigma_0$ is bounded by $C\varepsilon^\rho$ as well;

c) each torus $\Sigma_\varepsilon(\mathbb{T}^n \times \{\omega\})$, $\omega \in \Omega_\varepsilon$, is invariant for the flow of equation (8.7) and is filled with its quasiperiodic solutions $\mathfrak{h}(t)$ of the form $\mathfrak{h}(t;q,\omega) = \Sigma_\varepsilon(q + \omega't, \omega)$, where $\omega' = \omega'(\omega)$ and $|\omega' - \omega| + \mathrm{Lip}\,(\omega' - \omega) \le C\varepsilon^\rho$.

In statement b) of the theorem we view the difference $\Sigma_\varepsilon - \Sigma_0$ as a map, valued in the Hilbert space $\mathbb{R}^{2n} \times Y_d$.

Amplification. *Assertions b), c) hold with ρ replaced by one.*

Theorem 10. *If in Theorem 9 all the frequencies ν_j are real, then a solution $\mathfrak{h}(t)$ is linearly stable provided that the equation (8.7) linearised about this solution is uniformly well defined in the space $\mathbb{R}^{2n} \times Y_d$.*

For proofs these results with $d_1 \ge 1, \tilde{d} \le 0$ see [K,P] and with $d_1 > 1$ see [KK, K2].

8.4   Completion of the Main Theorem's proof (Step 4)

Now we apply Theorem 9 to equation (8.7) with Ω equal to a Borel subset Ω_l of the domain $\widetilde{\Omega}_l = \{\omega(r) \mid r \in \widetilde{R}_l\}$, $l = 1,\ldots,M$, which we construct below, and with $\gamma = \gamma_0/(MC_0)$, where the number C_0 will be be chosen later.

The assumptions 1)–3) hold with the constants from n through $d_1^1, d_1^2, \ldots$ the same as in Theorem 7, while the constants K and K_2 depend on γ. We take $j_2 = j_2(\gamma)$ and $M_1 = M_1(\gamma)$ as in Theorem 9 and consider all resonances as in the l.h.s. of (8.9). Since the system of Floquet exponents $\{\nu_j(r)\}$ is nondegenerate, then each resonance does not vanish identically. As these functions are analytic, we can find $K_3 = K_3(\gamma)$ and for every l can find a subset $\Omega_l \subset \widetilde{\Omega}_l$ such that $\mathrm{mes}\,(\widetilde{\Omega}_l \setminus \Omega_l) \le \gamma/M$ and (8.9) holds for all $\omega \in \Omega_l$.

For every l we apply Theorem 9 with γ as above to find the subset $\Omega_{l\varepsilon} \subset \Omega_l$, $\mathrm{mes}\,(\Omega_l \setminus \Omega_{l\varepsilon}) < \gamma_0$, and the map $\Sigma_{l\varepsilon} : \mathbb{T}^n \times \Omega_{l\varepsilon} \longrightarrow Q_\delta$.

Now we are in position to define the set $\widetilde{R}_\varepsilon \subset \widetilde{R}$ and the map $\Sigma^\varepsilon : \widetilde{R}_\varepsilon \times \mathbb{T}^n \longrightarrow Q_\delta$, claimed in Theorem 7. We set:

$$\widetilde{R}_\varepsilon = \bigcup_{l=1}^M \{r \in R_l \mid \omega(r) \in \Omega_{l\varepsilon}\}, \quad \Sigma^\varepsilon(r,\mathfrak{z}) = G \circ \mathrm{Shift}_p \circ \Sigma_{l\varepsilon}(q(\mathfrak{z}),\omega(r))$$

for r in R_l, where $(r,\mathfrak{z}) \mapsto (p,q)$ is the action-angle transformation from Step 1.

The set $\widetilde{R}_\varepsilon$ and the map Σ^ε satisfy all the claims of Theorem 7. Indeed,

$$\mathrm{mes}\,(R \setminus \widetilde{R}_\varepsilon) = \mathrm{mes}\,(\widetilde{R} \setminus \cup\tilde{R}) + \sum_{l=1}^M \mathrm{mes}\,\{r \in \widetilde{R}_l \mid \omega(r) \in \widetilde{\Omega}_l \setminus \Omega_l\}$$

$$+ \sum_{l=1}^M \mathrm{mes}\,\{r \in \widetilde{R}_l \mid \omega(r) \notin \Omega_{l\varepsilon}\}.$$

Denoting $\sup|\det \partial\omega/\partial r|^{-1}$ by $C(\gamma_0)$ we see that $\mathrm{mes}\,(R \setminus \widetilde{R}_\varepsilon)$ is bounded by $\gamma_0 + \gamma_0 + C(\gamma_0)M\gamma_0/(MC_0)$. This is smaller than $3\gamma_0$, if $C_0 = C_0(\gamma_0)$ is big enough. It means that we can choose $\gamma = \gamma(\varepsilon)$ in such a way that $\mathrm{mes}\,(R \setminus \widetilde{R}) \le 3\gamma_0$ goes to zero with ε and the assertion a) of Theorem 7 holds.

The tori $\Sigma^\varepsilon(\{r\} \times \mathbb{T}^n)$ are invariant for equation (8.3) and are filled with its quasiperiodic solutions of the form $\mathfrak{h}_\varepsilon(t)$, where $\omega_\varepsilon = \omega'(p(r))$.

The estimates for $\Sigma^\varepsilon - \Phi_0$ and $\omega_\varepsilon(r) - \omega(r)$ readily follows from the corresponding estimates in Theorem 9.

It remains to estimate Lipschitz constants of the differences as above. Let us take any two points $(r_1, \mathfrak{z}_1)$ and $(r_2, \mathfrak{z}_2)$ in $\widetilde{R}_\varepsilon \times \mathbb{T}^n$. If r_1 and r_2 belong to the same set $\widetilde{R}_l$, then the estimates for increments[17] of $\Sigma^\varepsilon - \Phi_0$ and $\omega_\varepsilon - \omega$ follow from the corresponding estimates for the increments of $\Sigma(0, q, 0)$ and $\omega' - \omega$ since the maps (8.16) are Lipschitz. If r_1 and r_2 belong to different sets $\widetilde{R}_l$, then $|r_1 - r_2| \ge C(\gamma) > 0$ and the increments of the differences divided by the increments of the arguments is bounded by $C_1\varepsilon^\rho/C(\gamma)$. Since we can choose the rate of decaying $\gamma(\varepsilon) \to 0$ to be as slow as we wish, then we can achieve $C_1\varepsilon^\rho/C(\gamma) \le C_2\varepsilon^{\tilde\rho}$, if we chose for ρ in Theorem 9 any number from the interval $(\tilde\rho, 1/3)$.

The last arguments also show that the estimates $|\omega' - \omega| \le C\varepsilon$ and $\mathrm{Lip}\,(\omega' - \omega) \le C\varepsilon$ imply that $|\omega_\varepsilon - \omega| \le C\varepsilon$ and $\mathrm{Lip}\,(\omega_\varepsilon - \omega) \le C_{\rho'}\varepsilon^{\rho'}$ for any $\rho' < 1$. It means that the Amplification to Theorem 9 implies the Amplification to Theorem 7.

Finally, since linearisation of the symplectomorphism which sends solutions $\mathfrak{h}(t)$ of (8.12) to solutions $\mathfrak{h}_\varepsilon(t)$ transforms solutions of the corresponding linearised equations, then Theorem 8 follows from Theorem 10.

9 Examples

9.1 Perturbed KdV equation

Let us consider a perturbed KdV equation

$$\dot{u} = \frac{1}{4}u_{xxx} + \frac{3}{2}uu_x + \varepsilon\frac{\partial}{\partial x}f_u'(u, x) =: V_\varepsilon(u)(x), \qquad (9.1)$$

under zero mean-value periodic boundary conditions. In (9.1) $f(u, x)$ is a C^d-smooth function $(d \ge 1)$, δ-analytic in u. Then the nonlinear part of the vector field V_ε defines an analytic map of order one:

$$H_0^d \longrightarrow H_0^{d-1}, \quad u \longmapsto \frac{3}{2}uu_x + \varepsilon\frac{\partial}{\partial x}f_u'(u, x).$$

The equation's hamiltonian is $\mathcal{H}_\varepsilon = \int_0^{2\pi} \left(\frac{1}{8}u'^2 - \frac{1}{4}u^3 - \varepsilon f(u(x), x)\right) dx$.

---

[17] i.e., the estimate $|(\omega_\varepsilon - \omega)(r_1) - (\omega_\varepsilon - \omega)(r_2)| \le C\varepsilon^\rho|r_1 - r_2|$, etc.

For $\varepsilon = 0$ this is the KdV equation and defined in (4.5) bounded part $\mathcal{T}^{2n}$ of any finite-gap manifold $\mathcal{T}_V^{2n}$ satisfies the assumptions i)-v) (see in Section 3). The linearised KdV equation has a system of Floquet solutions which is complete nonresonant (Section 6.2). The assumption 1) of Theorem 7 now holds with $d_1 = 3$, $d_1^1 = \cdots = 0$ and 2) holds since $d_H = \tilde{d} = 1$. We get:

Theorem 11. *For any $\rho < 1$ and for sufficiently small $\varepsilon > 0$, there exists a Borel subset R_ε^n of the cube $R^n = \{0 < r_j < K\}$ and a Lipschitz map $\Sigma^\varepsilon : R_\varepsilon^n \times \mathbb{T}^n \longrightarrow H_0^d(S^1)$, analytic in the second variable, such that:*

a) $\text{mes}_n(R^n \setminus R_\varepsilon^n) \longrightarrow 0$ as $\varepsilon \to 0$,

b) the map Σ^ε is ε^ρ-close to the map Φ_0, $\Phi_0(r, \mathfrak{z})(x) = G(\mathbf{V}x + \mathfrak{z}, r)$ (see (3.16')), also in the Lipschitz norm,

c) each torus $T_\varepsilon^n(r) = \Sigma^\varepsilon(\{r\} \times \mathbb{T}^n)$, $r \in R_\varepsilon^n$, is invariant for equation (9.1) and is filled with its linearly stable time-quasiperiodic solutions of the form $t \to \Sigma^\varepsilon(r, \mathfrak{z} + t\omega_\varepsilon(r))$, where the n-vector ω_ε is $C\varepsilon$-close to $\mathbf{W}(r)$.

To get the result we used Theorem 7, its Amplification and Theorem 8. The last theorem applies since the Floquet solutions of the linearised KdV equation have real exponents and since the linearised KdV equation is well posed.

The theorem implies that the union of all linearly stable time-quasiperiodic solutions becomes infinite-dimensional and dense in H_0^d asymptotically as $\varepsilon \to 0$:

Corollary. *The space H_0^d contains a subset Q_ε filled with linearly stable time-quasiperiodic solutions of (9.1) such that its Hausdorff dimension tends to infinity when $\varepsilon \to 0$ and for any fixed function $v \in H_0^d$ we have:*

$$\text{dist}_{H_0^d}(v, Q_\varepsilon) \longrightarrow 0 \quad as \quad \varepsilon \to 0. \tag{9.2}$$

Proof. We define Q_ε as a union of all non-empty sets $\Sigma^\varepsilon(R_\varepsilon^n \times \mathbb{T}^n) = \widetilde{\mathcal{T}}_\varepsilon^{2n}$, corresponding to all n-gap manifolds $\mathcal{T}^{2n}$, $n = 1, 2, \ldots$. By Proposition 4, the set $\widetilde{\mathcal{T}}_\varepsilon^{2n}$ has positive $2n$-dimensional Hausdorff measure when ε is small. Thus, $\dim_{\mathcal{H}} Q_\varepsilon \longrightarrow \infty$.

To prove (9.2) we note that for any $\mu > 0$ one can find $n \geq 1$ and an n-gap potential $u(x)$ such that $\|u - v\|_k \leq \mu$ (this is a famous result of V.A.Marchenko, see [Ma], Theorem 3.4.3 and [GT], p.27). Then u equals to $\Phi_0(r, \mathfrak{z})$ with some $r \in R$ and $\mathfrak{z} \in \mathbb{T}^n$. If ε is sufficiently small, then by the assertion a) of the theorem, there exists $r_1 \in R_\varepsilon$ such that $|r - r_1| \leq \mu$. Using b), we get that $\|u - \Sigma^\varepsilon(r_1, \mathfrak{z})\|_k \leq C\mu + \varepsilon^\rho$ and (9.2) follows since $\mu > 0$ can be chosen arbitrary small. $\square$

Another immediate consequence of the theorem is the observation that the Its–Matveev formula (4.4) with corrected frequency vector $\mathbf{W}$ "almost solves" the equation (9.1) for all t:

Corollary. *For any $r \in \mathbb{R}^n_+$ and any $\mathfrak{z} \in \mathbb{T}^n$ there exists an n-vector $\mathbf{W}_\varepsilon(r)$ and a solution $u_\varepsilon(t, x)$ of (9.1) in H_0^d such that*

$$\sup_t \|u_\varepsilon(t, \cdot) - 2\frac{\partial^2}{\partial x^2} \ln \theta(i(\mathbf{V} \cdot + \mathbf{W}_\varepsilon t + \mathfrak{z}); r)\|_d \longrightarrow 0 \quad as \quad \varepsilon \to 0.$$

Proof. For every $\varepsilon > 0$ let us choose any $r_\varepsilon \in R_\varepsilon$ such that $r_\varepsilon \to r$ as $\varepsilon \to 0$ and take $u_\varepsilon(t) = \Sigma^\varepsilon(r_\varepsilon, \mathfrak{z} + t\omega_\varepsilon(r))$. Then

$$\|u_\varepsilon(t) - \Phi_0(r, \mathfrak{z} + \omega_\varepsilon t)\|_k \le \|u_\varepsilon(t) - \Phi_0(r_\varepsilon, \mathfrak{z} + \omega_\varepsilon t)\|_k$$
$$+ \|\Phi_0(r_\varepsilon, \mathfrak{z} + \omega_\varepsilon t) - \Phi_0(r, \mathfrak{z} + \omega_\varepsilon t)\|_k = o(1) \quad \text{as } \varepsilon \to 0.$$

This implies the result since $\Phi_0(r, \mathfrak{z} + \omega_\varepsilon t)(x) = 2\frac{\partial^2}{\partial x^2} \ln \theta(i(\mathbf{V}x + \mathbf{W}_\varepsilon t + \mathfrak{z}); r)$, where $\mathbf{W}_\varepsilon = \omega_\varepsilon$. $\square$

An easy analysis of the first step in the proof of Theorem 9 (see [KK]) shows that the new frequency vector $\mathbf{W}_\varepsilon$ has the form $\mathbf{W}_\varepsilon(r) = \mathbf{W}(r) + \varepsilon \mathbf{W}_1(r) + O(\varepsilon^2)$, where components W_1^j of the n-vector $\mathbf{W}_1$ are obtained by averaging along the torus $T^n(r)$ [18] of the function $\left[G_*\left(\frac{\partial}{\partial p_j}\right)\right]\left(-\int_0^{2\pi} f(u, x)\, dx\right)$. Here $G : (p, q, y) \mapsto u(\cdot)$ is the normal form transformation from Theorem 11.

Therefore the assertion of the second Corollary can be viewed as an averaging theorem: for most r and for all $\mathfrak{z}$ the functions $2\frac{\partial^2}{\partial x^2} \ln \theta(i(\mathbf{V}x + \mathbf{W}_\varepsilon t + \mathfrak{z}); r)$ with $\mathbf{W}_\varepsilon = \mathbf{W}(r) + \varepsilon \mathbf{W}_1(r) + O(\varepsilon^2)$ approximate solutions of the perturbed KdV equation (9.1) for *all* t and x, where the n-vector $\mathbf{W}_1$ is obtained by the averaging described above. Here "for most r" means "for all r outside a set whose measure goes to zero with ε".

9.2   Higher KdV equations

Let us consider a perturbation of the l-th equation from the KdV-hierarchy:

$$\dot{u} = \frac{\partial}{\partial x}(\nabla_u \mathcal{H}_l + \varepsilon \nabla_u H_1), \tag{9.3}$$

where $\mathcal{H}_l(u) = K_l \int_0^{2\pi} \left(u^{(l)^2} + \langle \text{higher-order terms with} \le l - 1 \text{ derivatives}\rangle \right) dx$ and $H_1 = \int_0^{2\pi} f(x, u, \ldots, u^{(l-1)}) \, dx$. The function f is assumed to be C^d-smooth in $x, \ldots u^{(l-1)}$ and δ-analytic in $u, \ldots, u^{(l-1)}$. Since

$$\nabla_u H_1 = \sum_{j=0}^{l-1} (-1)^j \frac{\partial^j}{\partial x^j} f'_{u^{(j)}}(x, \ldots, u^{(l-1)}),$$

then arguing as in Example 1.1, we see that $\frac{\partial}{\partial x} \nabla_u H_1$ is an analytic map of order $2l - 1$.

---

[18] with respect to the measure $(2\pi)^{-n} dq = (2\pi)^{-n} d\mathfrak{z}$.

Let us take a bounded part $\mathcal{T}^{2n}$ of any n-gap manifold. It is invariant for the l-th KdV-equation (equal to $(9.3)_{\varepsilon=0}$). As an invariant subset, it satisfies the assumptions i)-iv). The linearised equation has a complete system of Floquet solutions (see in Section 6). Due to (4.6) this system is nonresonant.

Now Theorem 7 applies to equation (9.3) since the assumption 1) holds with $d_1 = 2l+1$, $d_1^1 = \cdots = 0$ (see (6.20)) and 2) holds with $d_H = \tilde{d} = 2l+1$.

We see that *most of n-gap solutions of the l-th KdV equation persist in the perturbed equation (9.3) with sufficiently small ε in the same sense as for the KdV equation.*

9.3 Perturbed SG equation

Since the system of finite-gap solutions (4.7), (4.8) of the SG equation under the even periodic boundary conditions satisfy the assumptions i)-iv) (Section 4) and the system of Floquet solutions (6.23) is complete nonresonant, then Theorem 7 applies to the perturbed equation,

$$\begin{cases} u_{tt} - u_{xx} + \sin u & = \varepsilon f(u,x), \\ u(t,x) \equiv u(t,x+2\pi) & \equiv u_0(-x), \end{cases} \tag{9.4}$$

where f is C^k-smooth function ($k \geq 1$), analytic in u. Accordingly, for any $n \geq 1$, most of finite-gap solutions (4.7), (4.8) persist as time-quasiperiodic solutions of (9.4). The persisted solutions are Lyapunov-stable if the open gaps for solutions (4.7) are sufficiently small. Otherwise, they in general are non-stable.

Since it is unknown if the finite-gap solutions of (SG) jointly are dence in a function space, then we do not know if the persisted solutions of (9.4) are asymptotically dense as $\varepsilon \to 0$ (cf. (9.2)).

9.4 KAM-persistence of lower-dimensional invariant tori of nonlinear finite-dimensional systems

Let $\mathbb{R}^{2N}$ be an euclidean space, given the usual symplectic structure, let $\mathcal{T}^{2N} = \bigcup_{r \in R} T_r^n$ be an analytic submanifold of $\mathbb{R}^{2N}$, diffeomorphic to $R \times \mathbb{T}^n$, $R \Subset \mathbb{R}^n$, and $H_1, \ldots, H_n$ be commuting hamiltonians, as in Proposition 3 (so they are defined and analytic in the vicinity of $\mathcal{T}^{2n}$ and each torus T_r^n is invariant for every hamiltonian vector field V_{H_j}).

Let us take any hamiltonian – say, H_1. Then the vector field $V_{H_1} \mid_{\mathcal{T}^{2n}}$ has the form $\sum \omega_l(r) \partial/\partial_{3l}$ and by Proposition 3 linearised equations have Floquet solutions with some frequencies $\nu_j(r)$, which are real analytic functions.

Applying Theorem 7 we get that:

Theorem 12. *Let us assume that the following analytic functions do not vanish identically:*

$$l \cdot \nu(r) + s \cdot \omega(r), \quad l \in \mathbb{Z}^{N-n}, \ 1 \leq |l| \leq 2; \ s \in \mathbb{Z}^n. \tag{9.5}$$

Let h be an analytic function, defined in the vicinity of $\mathcal{T}^{2n}$. Then most of the tori T_r^n persist as invariant n-tori of the perturbed Hamiltonian vector field $V_{H_1+\varepsilon h}$, $0 < \varepsilon \ll 1$, in the sense, specified in Theorem 7. The persisted tori are filled with quasiperiodic solutions with zero Lyapunov exponents.

This reduction of the Main Theorem is much easier than the Main Theorem itself. Its claim remains essentially true under weaker assumptions: it is sufficient to check that only functions (9.5) with $|l| = 1$ do not vanish identically, see [Bour] (we note that under this weaker assumption the claim about Lyapunov exponents is not true).

References

[A1]   Arnold V.I., *Mathematical methods in classical mechanics.* 3rd ed. Moscow, Nauka, 1974; English transl., 2nd ed., Springer-Verlag, Berlin, 1989

[Ad]   Adams R. S., *Sobolev spaces.* Academic Press, New York–London, 1975

[AG]   Arnold V.I., Givental A.B., *Symplectic geometry.* In Encycl. of Mathem. Scien., vol. 4, Springer-Verlag, Berlin, 1989

[BB]   Belokolos E.D., Bobenko A.I., Enolskii V.Z., Its A.R., Matveev V.B., *Algebro-geometric approach to nonlinear integrable equations.* Springer-Verlag, Berlin, 1994

[BiK1]   Bikbaev R.F., Kuksin S.B., *A periodic boundary value problem for the Sine–Gordon equation, small Hamiltonian perturbations of it, and KAM-deformations of finite-gap tori.* Algebra i Analiz **4:3** (1992) 42–78; English transl. in St.-Petersburg Math. J. **4** (1993) 439–468

[BiK2]   Bikbaev R.F., Kuksin S.B., *On the parametrization of finite-gap solutions by frequency vector and wave-number vector and a theorem of I.Krichever.* Lett. Math. Phys. **28** (1993) 115–122

[BKM]   Bättig D., Kappeler T., Mityagin B., *On the KdV equation: convergent Birkhoff normal form.* J. Funct. Anal. **140** (1996) 335–358

[BM]   Bochner S., Martin W.T., *Several complex variables.* University Press, Princeton, 1948

[BoK1]   Bobenko A.I., Kuksin S.B., *Finite-gap periodic solutions of the KdV equation are nondegenerate.* Phis. Lett. A **161:3** (1991) 274–276

[BoK2]   Bobenko A.I., Kuksin S.B., *The nonlinear Klein-Gordon equation on an interval as a perturbed Sine-Gordon equation.* Comment. Math. Helv. **70** (1995) 63–112

[Bour]   Bourgain J., *On Melnikov's persistency problem.* Math. Res. Lett. **4** (1997) 445–458

[BT]   Bikbaev R.F., Tarasov V.O., *An inhomogeneous boundary value problem on the half-axis and on an interval for the Sine-Gordon equation.* Algebra i Analiz **3:4** (1991) 78–92; English transl. in St.-Petersburg Math. J. **3** (1992)

[Ca]   Cartan H. *Formes différerentielles.* Hermann, Paris, 1967

[D]   Dubrovin B.A., *Theta functions and nonlinear equations.* Uspekhi Mat. Nauk **36:2** (1981) 11–80; English transl. in Russ. Math. Surv. **36** (1981)

[DMN] Dubrovin B.A., Matveev V.F., Novikov S.P., *Nonlinear equations of Korteweg-de Vries type, finite zone linear operators, and Abelian varieties.* Uspekhi Mat. Nauk **31:1** (1976) 55–136; English transl. in Russ. Math. Surv. **31:1** (1976)

[EFM] Ercolani, N., Forest, M.G., McLaughlin, D., *Geometry of the modulational instability.* Part 1. Preprint, University of Arizona, 1987

[Fal] Falconer K., *Fractal Geometry.* l John Wiley & Sons, 1990

[Go] Godbillon C., *Géométrie différentielle et mécanique analytique.* Hermann, Paris, 1969

[GR] Gunning R.C., Rossi H., *Analytic functions of several complex variables.* Prentice-Hall, Inc., Englewood Cliffs, N.J., 1965

[GS] Guillemin V., Sternberg H., *Geometric asymptotics.* Amer. Math. Soc., Providence, 1977

[GT] Garnet J., Trubowitz, E., *Gaps and bands on one dimensional Schrödinger operator* II. Comment. Math. Helvetici **62** (1987) 18–37

[HS] Halmos P., Sunder V. *Bounded integral operators on L^2-spaces.* Springer-Verlag, Berlin, 1978

[K] Kuksin S.B., *Nearly integrable infinite-dimensional Hamiltonian systems.* Lecture Notes Math. 1556. Springer-Verlag, Berlin, 1993

[KK] Kuksin S.B., *Analysis of hamiltonian PDEs.* Manuscript

[K1] Kuksin S.B., *Perturbation theory for quasiperiodic solutions of infinite-dimensional Hamiltonian systems, and its applications to the Korteweg-de Vries equation.* Matem. Sbornik **136(178):3** (1988); English transl in Math. USSR Sbornik **64** (1989) 397–413

[K2] Kuksin S.B., *A KAM-theorem for equations of the Korteweg–de Vries type.* Rev. Math. & Math. Phys. **10:3** (1998) 1–64

[K3] Kuksin S.B., *An infinitesimal Liouville–Arnold theorem as criterion of reducibility for variational Hamiltonian equations.* Chaos, Solitons and Fractals **2:2** (1992) 259-269

[Kap] Kappeler T., *Fibration of the phase space for the Korteweg–de Vries equation.* Ann. Inst. Fourier **41** (1991) 539–575

[Kat1] Kato T., *Quasi-linear equations of evolutions, with applications to partial differential equations.* Lecture Notes Math. **448**, 25–70, Springer-Verlag, Berlin, 1975

[Kat2] Kato T. *Perturbation theory for linear operators.* Springer-Verlag, Berlin, 1966

[LM] Lions J.-L., Magenes E., *Problémes aux limites non homogénes et applications.* Dunod, Paris, 1968; English transl. Springer-Verlag, Berlin, 1972

[Ma] Marčenko, V.A., *Sturm-Liouville operators and applications.* Naukova Dumka, Kiev, 1977 Russian; English transl. Birkhäuser, Basel, 1986

[McK] McKean H.P., *The sine-Gordon and sinh-Gordon equation on the circle.* Comm. Pure Appl. Math. **34** (1981) 197–257

[Mil] Milnor J., *Singular points of complex hypersurfaces.* University Press, Princeton, 1968

[Mo] Moser J., *Various aspects of integrable hamiltonian systems.* In Proc. CIME Conference, Bressansone, Italien, 1978. Progr. Math. 8. Birkhäuser, 1980

[MT] McKean H.P., Trubowitz E., *Hill's operator and hyperelliptic function theory in the presence of infinitely many branch points.* Comm. Pure Appl. Math. **29** (1976) 143–226

[N]   Novikov S.P., *A periodic problem for the Korteweg-de Vries equation, I*.
         Funktsional. Anal. i Prilozhen. **8:3** (1974) 54–66; English transl in Func-
         tional Anal. Appl. **8**, 236–246 (1974)

[Nek]   Nekhoroshev N.N., *The Poincaré-Lyapunov-Liouville-Arnold theorem*,
         Funktsional. Analiz i Priloz. **28:2** (1994); English transl in Functional
         Anal. Appl. **28**, 128–129 (1994)

[P]   Pöschel J., *A KAM-theorem for some nonlinear PDEs*. Ann. Scuola Norm.
         Sup. Pisa, Cl. Sci., IV Ser. 15

[Paz]   Pazy A., *Semigroups of linear operators and applications to partial differ-
         ential equations*. Springer-Verlag, Berlin, 1983

[PT]   Pöschel J., Trubowitz E., *Inverse spectral theory*. Academic Press, New
         York–London, 1987

[RS2]   Reed M., Simon B., *Methods of modern mathematical physics*. Vol. 2. Aca-
         demic Press, New York–London, 1975

[RS4]   Reed M., Simon B., *Methods of modern mathematical physics*, Vol. 4, Aca-
         demic Press, New York–London, 1978

[S]   Springer G., *Introduction to Riemann surfaces*, second ed.. Chelsea Publ.
         Company, New York, 1981

[ZM]   Zakharov V.E., Manakov S.V., Novikov S.P., Pitaevskij L.P., *Theory of
         solitons*. Nauka, Moscow, 1980, English transl. Plenum Press, New York,
         1984

Analytic linearization of circle diffeomorphisms

Jean-Christophe Yoccoz

1 Introduction

Let $\mathbb{T} = \mathbb{R}/\mathbb{Z}$ and $r \in \{0, +\infty, \omega\} \cup [1, +\infty)$. We denote $\text{Diff}_+^r(\mathbb{T})$ the group of homeomorphisms of $\mathbb{T}$ of class C^r and C^r-isotopic to the identity (if $r = 0$ it is the group of homeomorphisms of $\mathbb{T}$; if $r \geq 1$, $r \in \mathbb{R}^* \setminus \mathbb{N}$, it is the group of $C^{\lfloor r \rfloor}$ diffeomorphisms whose r-th derivative verifies a Hölder condition of exponent $r - \lfloor r \rfloor$; if $r = \omega$ it is the group of $\mathbb{R}$-analytic diffeomorphisms). We denote $D^r(\mathbb{T})$ the group of C^r-diffeomorphisms f of the real line such that $f - \text{id}$ is $\mathbb{Z}$-periodic.

One can embed $\mathbb{R}$ into $D^\omega(\mathbb{T})$ as the subgroup of *translations*

$$\alpha \in \mathbb{R} \mapsto R_\alpha \in D^\omega(\mathbb{T}) \quad \text{where} \quad R_\alpha : x \mapsto x + \alpha,$$

and one can embed the quotient $\mathbb{T} = \mathbb{R}/\mathbb{Z}$ into $\text{Diff}_+^\omega(\mathbb{T})$ as the subgroup of *rotations*

$$\alpha \in \mathbb{T} \mapsto R_\alpha \in \text{Diff}_+^\omega(\mathbb{T}) \quad \text{where} \quad R_\alpha : x \mapsto x + \alpha (\text{mod} 1).$$

One has

$$\text{Diff}_+^r(\mathbb{T}) \simeq D^r(\mathbb{T})/\{R_p, p \in \mathbb{Z}\}.$$

We consider the usual C^r topology on $D^r(\mathbb{T})$ and $\text{Diff}_+^r(\mathbb{T})$.

Rotations on $\mathbb{T}$ have very simple dynamics. Poincaré asked under which condition a given homeomorphism f of $\mathbb{T}$ is equivalent (in some sense, e.g.: measurably, topologically, smoothly, analytically, etc.) to some rotation. He also gave a first answer to this question, today known as Poincaré's classification theorem: if the rotation number of $\rho(f)$ is irrational then f is combinatorially conjugated to the rotation $R_{\rho(f)}$ (see Corollary 3.3).

In Section 3.3 we explain the Denjoy theory [De]. Fixing the rotation number to be irrational Denjoy proved that if we add regularity to the homeomorphism f (actually f is C^1 and Df has bounded variation) then f is topologically conjugated to the rotation $R_{\rho(f)}$. This theorem doesn't give us information about the regularity of the conjugacy h.

Denjoy proved also that the hypothesis $f \in C^{1+BV}$ cannot be relaxed too much, constructing examples of C^1 diffeomorphisms which are not topologically conjugated to a rotation (see Section 3.4). In [He1] there is a generalisation of this result: for each irrational α there exists a dense subset, in the space of $\mathcal{C}^1$ diffeomorphisms whose rotation number is α, such that every element of this subset is not topologically conjugated to a rotation.

The step to higher order differentiability for the conjugation h requires new techniques and additional hypotheses on the rotation number. The fundamental regularity criterion is the following:

Let $k \geq r \geq 1$, $f \in Diff_+^k(\mathbb{T})$. f is $\mathcal{C}^r$-conjugated to a rotation (then to the rotation $R_{\rho(f)}$) if and only if the family $(f^n)_{n \geq 0}$ is *bounded* in the $\mathcal{C}^r$ topology.

In [Ar1] it was proved that if the rotation number verifies a diophantine condition and if the analytic diffeomorphism f is close enough to a rotation, then the conjugation is analytic. At the same time examples of analytic diffeomorphisms, with irrational rotation number, for which the conjugation is not even absolutely continuous were given.

In [He1] Arnol'd's result was improved obtaining a global result: there exists a set $\mathcal{A} \subset (0,1)$, with full Lebesgue measure, such that if the rotation number of the $\mathcal{C}^\infty$ diffeomorphism belongs to $\mathcal{A}$, then it is $\mathcal{C}^\infty$ conjugated to a rotation. A similar result holds for finitely differentiable diffeomorphisms, but in this case the conjugacy is less regular: this phenomenon of *loss of differentiability* is typical of small divisors problems.

In the case of finite differentiability, the following result is due to Y. Katznelson and D. Ornstein [KO1,KO2], see also [Yo1,KS,SK].

Theorem 1.1 *Let f be a $\mathcal{C}^k$ circle diffeomorphism, $k \in \mathbb{R}$, $k > 2$, with rotation number α. Assume that α verifies the following diophantine condition: there exists $c > 0$ and $\tau \geq 0$ such that for all $n \in \mathbb{N}$, $n > 0$*

$$|e^{2\pi i n \alpha} - 1| > c|n|^{-\tau - 1}.$$

Then the homeomorphism h which conjugates f to R_α is of class $\mathcal{C}^{k-1-\tau-\epsilon}$ for all $\epsilon > 0$. If f is $\mathcal{C}^2$ and α is a number of constant type (i.e. $\tau = 0$) then h is absolutely continuous.

The previous result shows that the loss of differentiability is at most $1 + \tau + \epsilon$.

In Section 4 we will present our results in the analytic case. We distinguish between local and global situation. In the local case one must assume that the rotation number α belongs to the set $\mathcal{B}$ of Brjuno numbers (see Section 2.4). We have two statements: in the first one we assume that the diffeomorphism is analytic and univalent on some "big" complex strip $B_\Delta = \{z \in \mathbb{C} : |\Im m\, z| < \Delta\}$, in the second one that the diffeomorphism is analytic in some "thin" complex strip B_δ but it is very close to a rotation:

Theorem 1.2 *Let $f \in Diff_+^\omega(\mathbb{T})$, $\rho(f) = \alpha \in \mathcal{B}$ and assume that $\Delta > \frac{1}{2\pi} B(\alpha) + c$, where c is a universal constant. If f is analytic and univalent in B_Δ then there exists $h \in Diff_+^\omega(\mathbb{T})$, analytic in $B_{\Delta'}$, with $\Delta' = \Delta - \frac{1}{2\pi} B(\alpha) - c$, which conjugates f to R_α in $B_{\Delta'}$.*

Theorem 1.3 *Let $f \in \text{Diff}^\omega_+(\mathbb{T})$, $\rho(f) = \alpha \in \mathcal{B}$. Given $\delta > 0$ there exists $\epsilon_0(\alpha, \delta)$ such that if f is holomorphic in B_δ and there $|f(z) - z - \alpha| \leq \epsilon_0$ then there exists $h \in \text{Diff}^\omega_+(\mathbb{T})$ such that $f = h \circ R_\alpha \circ h^{-1}$ on $B_{\frac{\delta}{2}}$.*

Using the theorem (proven in [Yo2]) showing that the Brjuno condition is also necessary for the problem of linearization of germs of complex analytic diffeomorphisms of $(\mathbb{C}, 0)$, and following the construction described in [PM], which associates to each non-linearizable germ of diffeomorphism of $(\mathbb{C}, 0)$ a non-linearizable analytic diffeomorphism of $\mathbb{T}$ close to a rotation, one sees that the Brjuno condition is necessary also in our context.

In the global case we introduce a condition (condition $\mathcal{H}$, see Section 2.5) which is sufficient and in some sense also necessary, more exactly:

Theorem 1.4 *Let $f \in \text{Diff}^\omega_+(\mathbb{T})$ and let α be its rotation number. If $\alpha \in \mathcal{H}$, there exists $h \in \text{Diff}^\omega_+(\mathbb{T})$ such that $f = h \circ R_\alpha \circ h^{-1}$. Moreover if $\alpha \notin \mathcal{H}$ there exists $f \in \text{Diff}^\omega_+(\mathbb{T})$ with rotation number α, which is not C^ω-linearizable.*

Theorems 1.2 and 1.3 will be proved in Sections 4.2 and 4.3 respectively. The first part of Theorem 1.4 will be proved in Section 4.5, using an extension of Denjoy's theory described in Section 4.4. The last Section 4.6 is devoted to the construction of counterexamples showing that the assumption $\alpha \in \mathcal{H}$ in the global theorem is also necessary.

2 Arithmetics

2.1 Introduction

The aim of this section is to introduce some elementary but fundamental facts from the theory of approximation of irrational numbers by rationals.

The most important tool we will use here is the continued fraction expansion of a real number (see Section 2.2 and [HW]). As it will be clear in what follows, the action of the modular group $GL(2, \mathbb{Z})$ on $\overline{\mathbb{R}} = \mathbb{R} \cup \{\infty\}$ plays a fundamental role in the renormalization procedure we will use to study the linearization problem. To better understand this action one can introduce a fundamental domain $[0, 1)$ for one of the two generators (the translation T) and concentrate the attention to the inversion $x \mapsto 1/x$ restricted to $[0, 1)$. Then one translates back $1/x$ to the fundamental domain and iterates the process. This gives an expanding map on $[0, 1)$, the *Gauss' map A* (2.1). As A is more and more expanding as $x \to 0+$ one obtains a "microscope" and the symbolic dynamics of this map gives rise to the continued fraction algorithm. The convergents of the continued fraction expansion of an irrational number give its best rational approximations in a very precise sense (see Lemma 2.1).

The classical diophantine conditions are reviewed in Section 2.3. They can be recast in terms of the speed of growth of various quantities related to the continued fraction expansion. This opens the way to introducing two sets of "badly approximable" irrational numbers which include diophantine numbers and which we will prove being the optimal sets for the linearization problem (respectively local for the set $\mathcal{B}$, introduced in Section 2.4, and global for the set $\mathcal{H}$, introduced in Section 2.5).

2.2 Continued Fractions

We consider the map: $A : (0, 1) \to (0, 1)$ defined by:

$$A(x) = \frac{1}{x} - \left\lfloor \frac{1}{x} \right\rfloor \tag{2.1}$$

To each $\alpha \in \mathbb{R} \setminus \mathbb{Q}$, iterating A, we associate its continued fraction expansion as follows. Let $(\alpha_n)_{n \geq 0}$ be defined by:

$$\alpha_0 = \alpha - \lfloor \alpha \rfloor, \quad \alpha_n = A^n(\alpha_0) \quad \text{for } n > 0 \tag{2.2}$$

and $(a_n)_{n \geq 0}$ by

$$a_0 = \lfloor \alpha \rfloor \quad \alpha_{n-1}^{-1} = a_n + \alpha_n \quad \forall n \geq 1 \tag{2.3}$$

then

$$\alpha = a_0 + \cfrac{1}{a_1 + \cfrac{1}{\ddots + \cfrac{1}{a_n + \alpha_n}}}$$

or shortly $\alpha = [a_0, a_1, \ldots, a_n + \alpha_n]$.

The nth-convergent is defined by

$$\frac{p_n}{q_n} = [a_0, a_1, \ldots, a_n] = a_0 + \cfrac{1}{a_1 + \cfrac{1}{\ddots + \cfrac{1}{a_n}}} \tag{2.4}$$

It is easy to identify the numerator and denominator of the nth-convergent with the sequences $(p_n)_{n \geq -2}$ and $(q_n)_{n \geq -2}$, defined by:

$$p_{-2} = 0, \quad p_{-1} = 1, \quad \text{and} \quad p_n = a_n p_{n-1} + p_{n-2}, \quad \text{for all } n \geq 0$$
$$q_{-2} = 1, \quad q_{-1} = 0, \quad \text{and} \quad q_n = a_n q_{n-1} + q_{n-2}, \quad \text{for all } n \geq 0$$

Moreover $\alpha = \frac{p_n + p_{n-1}\alpha_n}{q_n + q_{n-1}\alpha_n}$ and $\alpha_n = -\frac{q_n \alpha - p_n}{q_{n-1}\alpha - p_{n-1}}$.

Let

$$\beta_{-1} = 1 \quad \beta_n = \prod_{j=0}^{n} \alpha_j \equiv (-1)^n (q_n\alpha - p_n) \quad \forall n \geq 0 \qquad (2.5)$$

then

$$q_{n+1}\beta_n + q_n\beta_{n+1} = 1$$
$$a_{n+1}\beta_n + \beta_{n+1} = \beta_{n-1}$$

and as $a_n \geq 1$ for all n

$$(q_n + q_{n+1})^{-1} < \beta_n < (q_{n+1})^{-1}. \qquad (2.6)$$

We then have the following standard results:

Lemma 2.1 (Best approximation) *Let $q \geq 0$, $p \in \mathbb{Z}$, $n \geq 0$; if $q\alpha - p$ is strictly between $q_n\alpha - p_n$ and $q_{n+1}\alpha - p_{n+1}$ then either $q \geq q_n + q_{n+1}$ or $p = q = 0$.*

We refer to [HW] for the elementary proof.

Corollary 2.2 *Consider the set $\{q\alpha - p; 0 \leq q < q_{n+1}, p \in \mathbb{Z}\}$, ordered as $\ldots < x_{-1} < 0 = x_0 < x_1 < \ldots$. Let $x_k = q\alpha - p$ be an element of this set. Then*

- *if n is even* $\begin{cases} x_{k+1} = x_k + \beta_n & \text{if } q < q_{n+1} - q_n \\ x_{k+1} = x_k + \beta_n + \beta_{n+1} & \text{if } q \geq q_{n+1} - q_n \end{cases}$

- *if n is odd* $\begin{cases} x_{k-1} = x_k + \beta_n & \text{if } q < q_{n+1} - q_n \\ x_{k-1} = x_k + \beta_n + \beta_{n+1} & \text{if } q \geq q_{n+1} - q_n \end{cases}$

Proof. We consider only the case n is even, the odd case being similar. We write $x_{k+1} = q'\alpha - p'$, $0 \leq q' < q_{n+1}$.

If $q < q_{n+1} - q_n$, $x_k + \beta_n = (q + q_n)\alpha - (p + p_n)$ belongs to the set. If we had $x_{k+1} < x_k + \beta_n$, both $q' \geq q$ (x_{k+1} being between x_k and $x_k + \beta_n$) and $q' \leq q$ ($x_k + \beta_n$ being between x_{k+1} and $x_{k+1} + \beta_n$) would be impossible by Lemma 2.1, a contradiction.

If $q \geq q_{n+1} - q_n$, $x_k + \beta_n + \beta_{n+1}$ belongs to the set. If we had $x_{k+1} < x_k + \beta_n + \beta_{n+1}$, as $q' \geq 0 > q - q_{n+1}$, x_{k+1} would be between $x_k = (x_k + \beta_{n+1}) - \beta_{n+1}$ and $(x_k + \beta_{n+1}) + \beta_n$ and we would have by Lemma 2.1 $q' \geq q + q_n \geq q_{n+1}$, again a contradiction. $\square$

Let us consider the standard action of $PGL(2,\mathbb{Z}) = GL(2,\mathbb{Z})/\{\pm I\}$ on $\overline{\mathbb{R}} = \mathbb{R} \cup \{\infty\}$ given by:

$$\text{for } g = \begin{pmatrix} a & b \\ c & d \end{pmatrix} \in GL(2,\mathbb{Z}), g \cdot \alpha = \frac{\alpha a + b}{\alpha c + d}. \qquad (2.7)$$

Then we have

Proposition 2.3 *Two irrational numbers α and α' belong to the same $GL\,(2,\mathbb{Z})$-orbit if and only if there exists $n \geq 0$ and $n' \geq 0$ such that $\alpha_n = \alpha'_{n'}$.*

2.3   Diophantine Conditions

Given $\tau \geq 0$, $\alpha \in \mathbb{R} \setminus \mathbb{Q}$ satisfies a *diophantine condition of order τ* (for short $\alpha \in CD\,(\tau)$) if there exists a constant $c > 0$ such that:

$$\left| \alpha - \frac{p}{q} \right| \geq \frac{c}{q^{2+\tau}} \quad \forall \frac{p}{q} \in \mathbb{Q}, q \geq 1. \tag{2.8}$$

Proposition 2.4 *Let $\tau \geq 0$ and $\alpha \in \mathbb{R} \setminus \mathbb{Q}$ then the following statements are equivalent*

1. $\alpha \in CD\,(\tau)$;
2. $q_{n+1} = \mathcal{O}\big(q_n^{\tau+1}\big)$;
3. $a_{n+1} = \mathcal{O}(q_n^{\tau})$;
4. $\alpha_{n+1}^{-1} = \mathcal{O}(\beta_n^{-\tau})$;
5. $\beta_{n+1}^{-1} = \mathcal{O}(\beta_n^{-\tau-1})$;

where $(\alpha_n)_{n\geq 0}$, $(p_n/q_n)_{n\geq 0}$, $(\beta_n)_{n\geq -1}$ and $(a_n)_{n\geq 1}$ are defined in Section 2.2.

The proof is elementary and is left as an exercise.

We say that α satisfies a *diophantine condition* if $\alpha \in CD = \cup_{\tau \geq 0} CD\,(\tau)$. For all τ, $CD\,(\tau)$ is invariant for the action of $GL\,(2,\mathbb{Z})$. For all $\tau > 0$, $CD\,(\tau)$ has full measure; on the other hand $(\mathbb{R} \setminus \mathbb{Q}) \setminus CD$ is a $\mathcal{G}_\delta$ dense subset of $\mathbb{R}$ and $CD\,(0)$ has zero measure.

2.4   Brjuno function and condition $\mathcal{B}$

Let $\alpha \in \mathbb{R} \setminus \mathbb{Q}$, we define (following [Yo2] and [MMY1]) the *Brjuno function* $B : \mathbb{R} \setminus \mathbb{Q} \to \mathbb{R}^+ \cup \{+\infty\}$

$$B\,(\alpha) = \sum_{n \geq 0} \beta_{n-1} \log \alpha_n^{-1} \tag{2.9}$$

where $(\alpha_n)_{n\geq 0}$ and $(\beta_n)_{n\geq -1}$ are defined in Section 2.2.

We say that $\alpha \in \mathbb{R} \setminus \mathbb{Q}$ satisfies the *condition $\mathcal{B}$* (or is a *Brjuno number*, for short $\alpha \in \mathcal{B}$) if $B\,(\alpha) < +\infty$.

Remark 2.5 As in [Yo2] and [MMY1] we extend the above definition to rational values setting $B\,(\alpha) = +\infty$ or $e^{-B(\alpha)} = 0$ for $\alpha \in \mathbb{Q}$.

The following proposition collects some properties of the Brjuno function (see [MMY1], Proposition 2.3, p. 277, for a proof).

Proposition 2.6 *The function B satisfies*

1. $B(\alpha) = B(\alpha + 1)$ for all $\alpha \in \mathbb{R}$;
2. for all $\alpha \in (0, 1)$ then $B(\alpha) = \log \alpha^{-1} + \alpha B(\alpha^{-1})$;
3. there exists a constant $C_1 > 0$ independent of α such that for all $\alpha \in \mathbb{R} \setminus \mathbb{Q}$

$$\left| B(\alpha) - \sum_{j \geq 0} \frac{\log q_{j+1}}{q_j} \right| \leq C_1 \tag{2.10}$$

where $(q_j)_{j \geq 0}$ are the denominators of the convergent of α defined in (2.4).

Remark 2.7 By 2. one has that $\mathcal{B}$ is $PGL(2, \mathbb{Z})$-invariant. Indeed the Brjuno function B is a $PGL(2, \mathbb{Z})$-cocycle (see [MMY2], Appendix 5).

Remark 2.8 Let $\alpha \in CD$, by Proposition 2.4 there exists $\tau \geq 0$ such that $q_{n+1} = \mathcal{O}(q_n^{\tau+1})$, thus $\frac{\log q_{n+1}}{q_n} \leq (1 + \tau) \frac{\log q_n}{q_n} + \frac{c}{q_n}$ and then $B(\alpha) < +\infty$. Therefore $CD \subset \mathcal{B}$ and the set $\mathcal{B}$ has full measure.

Remark 2.9 For positive τ we could define a condition $CD'(\tau)$ using the function $B^{(\tau)}(\alpha) = \sum_{j \geq 0} \beta_{j-1} \alpha_j^{-\frac{1}{\tau}}$ (note that the function $B^{(\tau)'}(\alpha) = \sum_{j \geq 0} \beta_{j-1}^{\tau} \alpha_j^{-1}$ would have the same features) as follows:

$$\alpha \in CD'(\tau) \text{ if and only if } B^{(\tau)}(\alpha) < +\infty.$$

Then the condition $CD'(\tau)$ is almost equivalent to condition $CD(\tau)$, in fact

$$CD'(\tau) \Rightarrow CD(\tau),$$
$$CD(\tau) \Rightarrow CD'(\tau') \quad \forall \tau' > \tau.$$

2.5 Condition $\mathcal{H}$

For $\alpha \in (0, 1)$, $x \in \mathbb{R}$, define

$$r_\alpha(x) = \begin{cases} \alpha^{-1}(x - \log \alpha^{-1} + 1) & \text{if } x \geq \log \alpha^{-1}, \\ e^x & \text{if } x \leq \log \alpha^{-1}. \end{cases} \tag{2.11}$$

Observe that r_α is of class $\mathcal{C}^1$ on $\mathbb{R}$, satisfying

$$r_\alpha(\log \alpha^{-1}) = Dr_\alpha(\log \alpha^{-1}) = \alpha^{-1}; \tag{2.12}$$
$$e^x \geq r_\alpha(x) \geq x + 1 \quad \text{for all} \quad x \in \mathbb{R}; \tag{2.13}$$
$$Dr_\alpha(x) \geq 1 \quad \text{for all} \quad x \geq 0. \tag{2.14}$$

For $\alpha \in \mathbb{R} \setminus \mathbb{Q}$, we now set, for $k > 0$

$$\Delta_k(\alpha) = r_{\alpha_{k-1}} \circ \cdots \circ r_{\alpha_0}(0).$$

An easy calculation gives

Lemma 2.10 *Let $\alpha \in \mathcal{B}$ and $k \geq 0$. Assume that $B(\alpha_k) \leq \Delta_k(\alpha)$. Then one has, for $l \geq k$*

$$\beta_{l-1}\Delta_l(\alpha) = \beta_{k-1}\Delta_k(\alpha) + \sum_{j=k}^{l-1} \beta_{j-1} - \sum_{j=k}^{l-1} \beta_{j-1}\log\alpha_j^{-1}$$

$$= \beta_{k-1}(\Delta_k(\alpha) - B(\alpha_k)) + \sum_{j=k}^{l-1} \beta_{j-1} + \beta_{l-1}B(\alpha_l)$$

$$\geq \beta_{l-1}B(\alpha_l) \geq \beta_{l-1}\log\alpha_l^{-1}.$$

Let us define, for $n \geq k \geq 0$:

$$\mathcal{H}_{k,n} = \{\alpha \in \mathcal{B}, B(\alpha_n) \leq \Delta_k(\alpha_{n-k})\},$$

and then

$$\mathcal{H} = \cap_{m\geq 0}(\cup_{k\geq 0}\mathcal{H}_{k,k+m}) = \{\alpha \in \mathcal{B}, \forall m \geq 0, \exists k \geq 0, B(\alpha_{m+k}) \leq \Delta_k(\alpha_m)\}$$

Proposition 2.11 1. *For $0 < k \leq n$, one has $\mathcal{H}_{k-1,n} \subset \mathcal{H}_{k,n} \subset \mathcal{H}_{k+1,n+1}$;*
2. *Let $\alpha \in \mathbb{R} \setminus \mathbb{Q}$. Then $\alpha \in \mathcal{H}$ (resp. $\alpha \in \mathcal{B}$) if and only if the orbit of α under $GL(2,\mathbb{Z})$ is contained in (resp. intersects) $\cup_{k\geq 0}\mathcal{H}_{k,k}$.*
3. *The subset $\mathcal{H}$ of $\mathbb{R}$ is a $F_{\sigma\delta}$ subset (i.e. a countable intersection of F_σ subsets).*

Proof. 1. From Lemma 2.10 we have $\mathcal{H}_{k,n} \subset \mathcal{H}_{k+1,n+1}$. Also $\Delta_1(\alpha_{n-k}) = 1$ and from (2.14) it follows that $\Delta_k(\alpha_{n-k}) \geq \Delta_{k-1}(\alpha_{n-k+1}) + 1$ and thus $\mathcal{H}_{k-1,n} \subset \mathcal{H}_{k,n}$.

2. Let $\alpha \in \mathbb{R} \setminus \mathbb{Q}$. Then $\alpha \in \mathcal{H}$ if and only if $\alpha_n \in \cup_{k\geq 0}\mathcal{H}_{k,k}$ for all $n \geq 0$. As the α_n's belong to the orbit of α under $GL(2,\mathbb{Z})$, we see that if the orbit of α is contained in $\cup_{k\geq 0}\mathcal{H}_{k,k}$ then $\alpha \in \mathcal{H}$. Conversely, if $\alpha \in \mathcal{H}$ and α' belongs to the orbit of α, by Proposition 2.3 there exist integers n and n' such that $\alpha_n = \alpha'_{n'}$. Then $\alpha'_{n'} \in \cup_{k\geq 0}\mathcal{H}_{k,k}$ and it follows from 1. that also $\alpha' \in \cup_{k\geq 0}\mathcal{H}_{k,k}$. On the other hand, we have by definition $\cup_{k\geq 0}\mathcal{H}_{k,k} \subset \mathcal{B}$, and $\mathcal{B}$ is invariant under $GL(2,\mathbb{Z})$. Conversely, if $\alpha \in \mathcal{B}$, let $k \geq B(\alpha)$ and $\alpha' \in \mathbb{R} \setminus \mathbb{Q}$ be such that $\alpha'_k = \alpha_0$. Then α' belongs to the orbit of α and we have by (2.13) $\Delta_k(\alpha') \geq k \geq B(\alpha'_k)$ hence $\alpha' \in \mathcal{H}_{k,k}$.

3. As B is lower semicontinuous, for every $n \geq k \geq 0$, $\mathcal{H}_{k,n}$ is a closed subset and the conclusion follows. □

Remark 2.12 One has the obvious inclusions $CD \subset \mathcal{H} \subset \mathcal{B}$. Moreover, by 2. $\mathcal{H}$ is $GL(2,\mathbb{Z})$-invariant.

Example 2.13 Consider the compact subset $K \subset [0,1]$ whose elements are irrationals α such that

$$a_1 = 1, \quad a_{i+1} \leq \exp(2a_i) - 1.$$

Let $\alpha \in K$, $j \geq 0$; denote by $\tilde{p}_n/\tilde{q}_n$ the convergents of α_j. For $n > 0$, we have

$$\left(\prod_{0 \leq l < n} \alpha_{j+l} \right) \log \alpha_{j+n}^{-1} \leq \tilde{q}_n^{-1} \log(a_{j+n+1} + 1) \leq 2\tilde{q}_n^{-1} a_{j+n} \leq 2\tilde{q}_{n-1}^{-1};$$

it follows that $\alpha_j \in \mathcal{B}$ and the series for the Brjuno function $B(\alpha_j)$ converge uniformly on K; thus each function $\alpha \mapsto B(\alpha_j)$ is continuous on K. Moreover

$$\log a_{j+1} \leq B(\alpha_j) \leq \log(a_{j+1} + 1) + 8$$

as $\sum_{n \geq 0} q_n^{-1} \leq 4$ for every irrational α.

We next check that both $\mathcal{H}$ and $\mathbb{R} \setminus \mathcal{H}$ intersect K along a dense subset.

Indeed, if $a_i = 1$ for large i, α belongs to the $GL(2, \mathbb{Z})$ orbit of the golden mean, and thus to $\mathcal{H}$.

On the other hand, assume that for $i \geq i_0$ one has $a_{i+1} \geq \exp(a_i)$. Then, for $i \geq i_0$, we have

$$a_i \leq \log a_{i+1} \leq \log \alpha_i^{-1} \leq B(\alpha_i) \leq 3a_i + 4,$$

and thus, starting from $\Delta_0(\alpha_{i_0}) = 0 < a_{i_0}$, we have for $k \geq 0$ by induction

$$\Delta_k(\alpha_{i_0}) < a_{i_0+k} \leq B(\alpha_{i_0+k})$$

which means that $\alpha \notin \cup_{k \geq 0} \mathcal{H}_{k, i_0+k}$.

Finally, we observe that $\mathcal{H} \cap K$ is a G_δ-subset of K. Indeed, it follows from the definition of $\mathcal{H}$ and Lemma 2.10 that $\alpha \in \mathcal{H}$ if and only if for all $m \geq 0$ there exists $k \geq 0$ such that $B(\alpha_{m+k}) < \Delta_k(\alpha_m)$; this last condition is, for fixed m and k, open in K ($\alpha \mapsto B(\alpha_{m+k})$ being continuous) and this proves our claim.

We conclude that $\mathcal{H}$ is not a F_σ subset of $\mathbb{R}$; this would imply that both $\mathcal{H} \cap K$ and $(\mathbb{R} \setminus \mathcal{H}) \cap K$ are G_δ-dense subsets of the Baire space K.

Let $\mathcal{L}$ be the function defined inductively for $x \geq 0$ by

$$\mathcal{L}(x) = \begin{cases} x & \text{if } 0 \leq x \leq 1 \\ \mathcal{L}(\log x) + 1 & \text{if } x \geq 1. \end{cases}$$

It is a C^1-diffeomorphism of $[0, +\infty)$. For $0 \leq t \leq 1$, we consider the C^1-diffeomorphism from $[0, +\infty)$ onto $[t, +\infty)$ satisfying

$$\mathcal{L}(E_t(x)) - \mathcal{L}(x) = t, \quad \forall x \geq 0.$$

Lemma 2.14 1. *The map $x \mapsto E_t(x) - x$ is convex non-decreasing on $[0, +\infty)$.*
2. *For any $\alpha \in [0, 1]$, the map $x \mapsto \mathcal{L}(r_\alpha(x)) - \mathcal{L}(x)$ is non-increasing on $[0, +\infty)$.*

Proof. 1. For $0 \leq x \leq 1 - t$, we have $E_t(x) = x + t$, and for $1 - t \leq x \leq 1$, we have $E_t(x) = e^{x+t-1}$; thus $E_t(x) - x$ is convex non-decreasing on $[0,1]$. Next, from the commutation relation

$$E_t(e^x) = e^{E_t(x)}, \quad \forall x \geq 0$$

we get

$$\log DE_t(e^x) = \log DE_t(x) + E_t(x) - x$$

which shows inductively that $E_t(x) - x$ is convex non-decreasing on all of $[0, +\infty)$.

2. For $0 \leq x \leq \log \alpha^{-1}$, we have $r_\alpha(x) = e^x$ hence $\mathcal{L}(r_\alpha(x)) - \mathcal{L}(x) = 1$. Let $x > \log \alpha^{-1}$; as $x < r_\alpha(x) < e^x$, the number $t = \mathcal{L}(r_\alpha(x)) - \mathcal{L}(x)$ belongs to $(0,1)$. Consider the map $w \mapsto E_t(w) - r_\alpha(w)$ for $w \geq \log \alpha^{-1}$. It is convex, negative at $\log \alpha^{-1}$, and vanishes at x; therefore the derivative at x is positive, and thus we must have $D(\mathcal{L} \circ r_\alpha - \mathcal{L})(x) < 0$. $\square$

Lemma 2.15 *Let $\alpha \in \mathcal{B}$.*

1. *For any $k \geq 0$, we have $B(\alpha_{k+1}) \leq \exp(B(\alpha_k) - 1)$.*
2. *If $\alpha \notin \cup_{k \geq 0} \mathcal{H}_{k,k}$, the positive sequence $\mathcal{L}(B(\alpha_k)) - \mathcal{L}(\Delta_k(\alpha_k))$ is decreasing.*

Proof. 1. The assertion follows immediately from the relation

$$B(\alpha_{k+1}) = e^u(B(\alpha_k) - u), \quad u = \log \alpha_k^{-1}$$

maximizing over u.

2. From the first assertion we have always

$$B(\alpha_{k+1}) \leq r_{\alpha_k}(B(\alpha_k) - 1) \leq r_{\alpha_k}(B(\alpha_k)).$$

Therefore the second assertion of Lemma 2.15 follows from the second assertion of Lemma 2.14. $\square$

We now consider an irrational number $\alpha \in \mathcal{B}$. Clearly, if $\alpha \notin \mathcal{H}$, we must have

$$\lim_{k \to \infty} B(\alpha_k) = +\infty.$$

Fix $c_0 > 0$ large enough such that, for any irrational $\tilde{\alpha}$,

$$\sum_{k \geq 0} \tilde{\beta}_{k-1}^{1/2} \leq c_0.$$

We define, starting with $k_0 = 0$, an increasing sequence of indices k_i setting

$$k_{i+1} = \inf\{k > k_i, \quad \log \alpha_k^{-1} \geq c_0^{-1} \left(\frac{\beta_{k_i}}{\beta_{k-1}}\right)^{1/2} B(\alpha_{k_i+1})\}.$$

The set on the right-hand side is indeed nonempty: otherwise, as

$$B(\alpha_{k_i+1}) = \sum_{k > k_i} \frac{\beta_{k-1}}{\beta_{k_i}} \log \alpha_k^{-1}$$

we would have

$$B(\alpha_{k_i+1}) < c_0^{-1} \left(\sum_{k > k_i} \left(\frac{\beta_{k-1}}{\beta_{k_i}} \right)^{1/2} \right) B(\alpha_{k_i+1}) \leq B(\alpha_{k_i+1}).$$

Lemma 2.16 *For* $\alpha \in B$ *and* k_i *as above we have*

$$\lim_{i \to +\infty} \mathcal{L}(B(\alpha_{k_i})) - \mathcal{L}(\log \alpha_{k_i}^{-1}) = 0.$$

If $\alpha \notin \mathcal{H}$

$$\lim_{i \to +\infty} \mathcal{L}(B(\alpha_{k_i+1})) - \mathcal{L}(B(\alpha_{k_i})) = 1.$$

Proof. For the first limit, we note that

$$\frac{\beta_{k_i}}{\beta_{k_{i+1}-1}} B(\alpha_{k_i+1}) \geq B(\alpha_{k_i+1}) \geq \log \alpha_{k_{i+1}}^{-1} \geq c_0^{-1} \left(\frac{\beta_{k_i}}{\beta_{k_{i+1}-1}} \right)^{1/2} B(\alpha_{k_i+1})$$

$$(2.15)$$

But it is easily checked that

$$\lim_{x \geq 1, y \geq 1, xy \to \infty} \mathcal{L}(xy) - \mathcal{L}(c_0^{-1} x^{1/2} y) = 0$$

from which the assertion follows.

The second part of the Lemma is then an immmediate consequence of the second part of Lemma 2.15: as $\alpha \notin \mathcal{H}$ the sequence $\mathcal{L}(B(\alpha_k)) - \mathcal{L}(\Delta_k(\alpha))$ is positive decreasing; setting $\varepsilon_i = \mathcal{L}(B(\alpha_{k_i})) - \mathcal{L}(\log \alpha_{k_i}^{-1})$, we have that, if $\mathcal{L}(\Delta_{k_i}(\alpha)) \leq \mathcal{L}(B(\alpha_{k_i})) - \varepsilon_i$, then

$$\mathcal{L}(\Delta_{k_i+1}(\alpha)) = \mathcal{L}(\Delta_{k_i}(\alpha)) + 1 \leq \mathcal{L}(B(\alpha_{k_i+1})).$$

This forces the second assertion of the Lemma. □

We now consider the two relations (2.15) and

$$\frac{1}{\beta_{k_{i+1}-1}} = \frac{\beta_{k_i}}{\beta_{k_{i+1}-1}} \frac{1}{\beta_{k_i-1}} \frac{1}{\alpha_{k_i}}.$$

We get

$$\lim \left[\mathcal{L}(\beta_{k_{i+1}-1}^{-1}) - \max \left(\mathcal{L}(\beta_{k_i-1}^{-1}), \mathcal{L}(\alpha_{k_i}^{-1}), \mathcal{L}\left(\frac{\beta_{k_i}}{\beta_{k_{i+1}-1}} \right) \right) \right] = 0$$

and

$$\lim\left[\mathcal{L}(B(\alpha_{k_{i+1}})) - \max\left(\mathcal{L}\left(\frac{\beta_{k_i}}{\beta_{k_{i+1}-1}}\right), \mathcal{L}(B(\alpha_{k_{i+1}}))\right)\right] = 0$$

which, with Lemma 2.16, transform to

$$0 = \lim\left[\mathcal{L}(\beta_{k_{i+1}-1}^{-1}) - \max\left(\mathcal{L}(\beta_{k_i-1}^{-1}), \mathcal{L}(\log\alpha_{k_i}^{-1}) + 1, \mathcal{L}\left(\frac{\beta_{k_i}}{\beta_{k_{i+1}-1}}\right)\right)\right]$$

$$= \lim\left[\mathcal{L}(\log\alpha_{k_{i+1}}^{-1}) - \max\left(\mathcal{L}\left(\frac{\beta_{k_i}}{\beta_{k_{i+1}-1}}\right), \mathcal{L}(\log\alpha_{k_i}^{-1}) + 1\right)\right].$$

We claim that

Proposition 2.17 *We have* $\lim_{i\to+\infty}\mathcal{L}(\log\alpha_{k_{i+1}}^{-1}) - \mathcal{L}(\beta_{k_{i+1}-1}^{-1}) = 0$.

Proof. If not we would have

$$\mathcal{L}(\beta_{k_i-1}^{-1}) > \mathcal{L}(\log\alpha_{k_i}^{-1}) + 1$$

for infinitely many i's. But setting

$$\varepsilon_i = \mathcal{L}(\beta_{k_{i+1}-1}^{-1}) - \max\left(\mathcal{L}(\beta_{k_i-1}^{-1}), \mathcal{L}(\log\alpha_{k_i}^{-1}) + 1, \mathcal{L}\left(\frac{\beta_{k_i}}{\beta_{k_{i+1}-1}}\right)\right)$$

$$\varepsilon_i' = \mathcal{L}(\log\alpha_{k_{i+1}}^{-1}) - \max\left(\mathcal{L}\left(\frac{\beta_{k_i}}{\beta_{k_{i+1}-1}}\right), \mathcal{L}(\log\alpha_{k_i}^{-1}) + 1\right)$$

$$\eta_i = \mathcal{L}(\beta_{k_i-1}^{-1}) - \mathcal{L}(\log\alpha_{k_i}^{-1})$$

we have that if $\eta_{i+1} > \varepsilon_i - \varepsilon_i'$ then

$$\max\left(\mathcal{L}(\beta_{k_i-1}^{-1}), \mathcal{L}(\log\alpha_{k_i}^{-1}) + 1, \mathcal{L}\left(\frac{\beta_{k_i}}{\beta_{k_{i+1}-1}}\right)\right) = \mathcal{L}(\beta_{k_i-1}^{-1})$$

and

$$\eta_{i+1} = \varepsilon_i - \varepsilon_i' + \mathcal{L}(\beta_{k_i-1}^{-1}) - \max\left(\mathcal{L}\left(\frac{\beta_{k_i}}{\beta_{k_{i+1}-1}}\right), \mathcal{L}(\log\alpha_{k_i}^{-1}) + 1\right)$$

$$\leq \varepsilon_i - \varepsilon_i' + \eta_i - 1. \qquad \square$$

Corollary 2.18 *Let* $\alpha \notin \mathcal{H}$. *Then for any* $t < 1$ *there are infinitely many rationals such that* $|\alpha - p/q| \leq (E_t(q))^{-1}$.

2.6 $\mathbb{Z}^2$-actions by translations and continued fractions

Let (e_{-1}, e_0) be the canonical basis of $\mathbb{Z}^2$. We consider the following action $\mathbb{Z}^2 \overset{\pi}{\hookrightarrow} \text{Homeo}_+(\mathbb{R})$: to (e_{-1}, e_0) we associate two commuting homeomorphisms by

$$\begin{cases} \pi(e_{-1}) & : x \mapsto x - 1 = T(x) \\ \pi(e_0) & : x \mapsto x + \alpha = f(x) \end{cases} \tag{2.16}$$

We perform the sequence of changes of basis of $\mathbb{Z}^2$ given by

$$\begin{pmatrix} e_{n-1} \\ e_n \end{pmatrix} = M_n \begin{pmatrix} e_{-1} \\ e_0 \end{pmatrix} \quad , \quad \text{where} \quad M_n = \begin{pmatrix} p_{n-1} & q_{n-1} \\ p_n & q_n \end{pmatrix},$$

thus

$$\begin{pmatrix} e_n \\ e_{n+1} \end{pmatrix} = A_{n+1} \begin{pmatrix} e_{n-1} \\ e_n \end{pmatrix} \quad , \quad \text{where} \quad A_{n+1} = \begin{pmatrix} 0 & 1 \\ 1 & a_{n+1} \end{pmatrix}.$$

The homeomorphisms corresponding to the new basis (e_{n-1}, e_n) will be $(\pi(e_{n-1}), \pi(e_n))$ where

$$\pi(e_n): \quad x \mapsto x + (-1)^n \beta_n.$$

We can now rewrite Lemma 2.1 (from now on we will use the multiplicative notation for $\mathbb{Z}^k$-actions):

Lemma 2.19 (Best approximation 2nd version) *Let $p, q \in \mathbb{Z}$, if $q\alpha - p$ is strictly between $e_n \cdot 0$ and $e_{n-1} \cdot 0$ then either $|q| \geq q_n + q_{n-1}$ or $p = q = 0$.*

3   The C^r theory for $r \geq 0$

3.1   The C^0 theory

We consider an action of $\mathbb{Z}^k$ on $\mathbb{R}$, i.e. an homomorphism $\pi: \mathbb{Z}^k \to \text{Homeo}_+(\mathbb{R})$. We will assume that the action has no common fixed point. We define

$$Z^+ = \{e \in \mathbb{Z}^k : e \cdot x > x \quad \forall x \in \mathbb{R}\};$$

we have then

$$-Z^+ = \{e \in \mathbb{Z}^k : e \cdot x < x \quad \forall x \in \mathbb{R}\}.$$

Proposition 3.1 *The subset Z^+ of $\mathbb{Z}^k$ has the following properties:*

1. *$Z^+ \neq \emptyset$;*
2. *$-Z^+ \cap Z^+ = \emptyset$;*
3. *if $e' \notin -Z^+$ and $e \in Z^+$ then $e + e' \in Z^+$;*
4. *for any $r \geq 1$, $e \in Z^+$ if and only if $re \in Z^+$;*
5. *for any $e \in Z^+$, $e' \in \mathbb{Z}^k$ there exists n such that $e' + ne \in Z^+$.*

Proof. 1. We prove by induction on k that if $Z^+ = \emptyset$ the action has a common fixed point. Write $e_1, \ldots, e_k$ for the canonical basis of $\mathbb{Z}^k$. When $k = 1$, e_1 must have a fixed point if neither e_1 nor $-e_1$ belong to Z^+. From the induction hypothesis, we know that the closed set F of points fixed by $e_1, \ldots, e_{k-1}$ is non empty; also e_k has a fixed point x_0. If $x_0 \in F$, we are done; otherwise, as e_k commutes with e_i, $i < k$, the set F is e_k-invariant, and the component of $\mathbb{R} \setminus F$ containing x_0 is fixed by e_k. But then an endpoint of this component in F is fixed by e_k.

2. Obvious

3. By assumption, there exists x_0 such that $e' \cdot x_0 \geq x_0$. As $e \in Z^+$, and thus has no fixed point, there exists for any $x \in \mathbb{R}$ an integer $n \in \mathbb{Z}$ such that $ne \cdot x_0 \leq x < (n+1)e \cdot x_0$. Then

$$(e + e') \cdot x \geq [(n+1)e + e'] \cdot x_0 \geq (n+1)e \cdot x_0 > x.$$

4. Obvious

5. Let $x_0 \in \mathbb{R}$; as above there exists $n \in \mathbb{Z}$ such that $(n-1)e \cdot (e' \cdot x_0) \geq x_0$. Then $(n-1)e + e' \notin -Z^+$ and $ne + e' \in Z^+$ by 3. $\square$

Remark 3.2 A subset $Z^+ \subset \mathbb{Z}^k$ satisfies the properties 1–5 of Proposition 3.1 if and only if there exists a non zero linear form $\rho : \mathbb{Z}^k \to \mathbb{R}$ such that

$$Z^+ = \{e \in \mathbb{Z}^k, \rho(e) > 0\},$$

the form ρ being then uniquely determined up to a positive scalar multiple.

We leave the proof as an exercise.

The form ρ may be defined as follows. Choose $e_0 \in Z^+$ and define, for $e \in \mathbb{Z}^k$

$$N^+(e) = \inf\{n \in \mathbb{Z}, ne_0 - e \in Z^+\}$$
$$N^-(e) = \sup\{n \in \mathbb{Z}, ne_0 - e \in -Z^+\}$$

Then

$$\rho(e) = \lim_{k \to \infty} \frac{N^+(ke)}{k} = \lim_{k \to \infty} \frac{N^-(ke)}{k}. \tag{3.1}$$

Fix such a linear form ρ and consider the action by translations

$$\pi_\rho(e) \cdot x = x + \rho(e), \quad \text{for } e \in \mathbb{Z}^k, x \in \mathbb{R}.$$

Given $x_0 \in \mathbb{R}$, ρ vanishes on the π-stabilizer of x_0; this allows to define a map h from the π-orbit of x_0 onto $\rho(\mathbb{Z}^k)$ by $h(e \cdot x_0) = \rho(e)$.

This map is immediately seen to be order preserving. When the $\mathbb{Z}$-rank of ρ is > 1, $\rho(\mathbb{Z}^k)$ is dense in $\mathbb{R}$, and h has a unique order preserving extension to $\mathbb{R}$, which is automatically continuous and surjective and still a semi-conjugacy from π to π_ρ. We thus have

Corollary 3.3 (Poincaré) *When the $\mathbb{Z}$-rank of ρ is > 1, there is an order preserving surjective continuous semi-conjugacy h from π to π_ρ:*

$$h(e \cdot x) = h(x) + \rho(e), \forall x \in \mathbb{R}, \forall e \in \mathbb{Z}^k$$

uniquely determined up to an additive constant.

3.2   Equicontinuity and topological conjugacy

Let F be an element of Diff^r_+ ($\mathbb{T}$) and $f \in D^r(\mathbb{T})$ a lift of F. To such an F we can associate the $\mathbb{Z}^2$-action π defined as follows: to the canonical basis of $\mathbb{Z}^2$ we associate

$$\begin{cases} f_{-1} & : x \mapsto x - 1 = T(x), \\ f_0 & : x \mapsto f(x). \end{cases} \tag{3.2}$$

Then $\rho(\pi) = (-1, \alpha)$ and the number $\alpha \in \mathbb{R}$ is called the *rotation number* of f (or associated to the $\mathbb{Z}^2$-action π; we will also write $\alpha = \rho(f)$ and we define the rotation number of F being $\rho(F) = \alpha(\mathrm{mod}1)$).

Poincaré's result (Corollary 3.3) can be recast in the following form:

Proposition 3.4 (Poincaré) *Let $f \in D^0(\mathbb{T})$, $\alpha = \rho(f) \in \mathbb{R} \setminus \mathbb{Q}$. There exists $h : \mathbb{R} \to \mathbb{R}$ continuous, monotone non-decreasing such that $h - \mathrm{id}$ is $\mathbb{Z}$-periodic and $h \circ f = R_\alpha \circ h$.*

Remark 3.5 By Proposition 3.4 $\mu = Dh$ is an F-invariant probability measure on $\mathbb{T}$ and h is a homeomorphism if and only if the support of μ is $\mathbb{T}^1$.

Let $s \in \{0, +\infty, \omega\} \cup [1, +\infty), s \leq r$. The fundamental criterion of C^s conjugacy of a diffeomorphism F to the rotation R_α, where $\alpha = \rho(F)$, has been stated in the Introduction. Here, following [He1] (Section II.9) we will prove a criterion that guarantees that F is topologically conjugated to R_α. For the general case we refer to [He1] (Chapitre IV).

Proposition 3.6 *Let $F \in \mathrm{Diff}_+(\mathbb{T})$, $\alpha = \rho(F)$. F is C^0 conjugated to R_α if and only if the family $(F^{n_i})_{i \in \mathbb{N}}$ is equicontinuous for some $n_i \to \infty$.*

Proof. The condition is clearly necessary: here we assume that $\alpha \in \mathbb{R} \setminus \mathbb{Q}$ and we prove that it is also sufficient. We leave to the reader the case $\alpha \in \mathbb{Q}$.

Let μ be an F-invariant probability measure. By Remark 3.5, F is topologically conjugated to R_α if and only if $\mathrm{supp}\mu = \mathbb{T}^1$. If one has $K = \mathrm{supp}\mu \neq \mathbb{T}^1$, since K is F-invariant, $\mathbb{T}^1 \setminus K$ is open and F-invariant. Let J be a connected component of $\mathbb{T}^1 \setminus K$. Then the intervals $(F^n(J))_{n \in \mathbb{Z}}$ are pairwise disjoint connected components of $\mathbb{T}^1 \setminus K$. Thus $|F^n(J)| \to 0$ as $n \to \pm\infty$, but this contradicts the equicontinuity of $(F^n)_{n \in \mathbb{N}}$. $\square$

3.3 The Denjoy theory.

Let F be an element of $Diff_+^2(\mathbb{T})$ and f a lift of F. Set $k = 2$ and consider the action $\mathbb{Z}^2 \xrightarrow{\pi} Diff_+^2(\mathbb{R})$ defined as in (3.2). We assume that the rotation number $\alpha \in \mathbb{R} \setminus \mathbb{Q} \cap (0,1)$. As usual $\left(\frac{p_n}{q_n}\right)_{n \geq 0}$ will be the convergents of α and we can define a new basis of $\mathbb{Z}^2$ by (f_{n-1}, f_n) where $f_n = f^{q_n} T^{p_n}$.

The goal of this section is to prove Denjoy's Theorem (Corollary 3.13): if the action is by C^2 diffeomorphisms then the semiconjugacy h of Corollary 3.3 is indeed a *conjugacy* (there is no interval on which h is constant). The fundamental step will be Corollary 3.12 which allows to obtain a bound for $|Df_n|_{C^0}$.

The following lemma is fundamental for all the theory, in fact it gives control of the reciprocal position of the iterates of a fixed interval.

Lemma 3.7 *Let* $I_n(x) = [x, f_n(x)]$ *and* $J_n(x) = I_n(x) \cup I_n\left(f_n^{-1}(x)\right) = [f_n^{-1}(x), f_n(x)]$. *Then*

1. *for* $0 \leq j < q_{n+1}$ *and* $k \in \mathbb{Z}$ *the intervals* $f^j T^k(I_n(x))$ *have disjoint interiors;*
2. *for* $0 \leq j < q_{n+1}$ *and* $k \in \mathbb{Z}$ *the intervals* $f^j T^k(J_n(x))$ *cover* $\mathbb{R}$ *(at most twice).*

Proof. We assume for instance that n is even. Since the two statements are invariant under conjugacy, we can assume that f is the translation by α and $x = 0$. With the notations of Corollary 2.2, write $f^j T^k(0) = x_l$; then $f^j T^k(I_n(0)) = (x_l, x_l + \beta_n)$ and $f^j T^k(J_n(0)) = (x_l - \beta_n, x_l + \beta_n)$. Therefore both statements in the Lemma follow from Corollary 2.2. □

Remark 3.8 We note that in what follows the hypothesis C^2 can be replaced by $f \in C^{1+BV}$, i.e. f is C^1 and Df has bounded variation. In fact we will always need control of variation of the function $\log Df$ (or of something similar) when Df is positive[1].

The following proposition shows that using Lemma 3.7 we can bound the variation of $\log Df^j$ on the intervals $I_n(x)$ with the total variation of $\log Df$ on $[0,1]$. Using Proposition 3.9 we then prove Corollaries 3.10 and 3.11, which, together with Corollary 3.12, imply Denjoy's Theorem (Corollary 3.13).

Proposition 3.9 *Let* $f \in C^2$ *then for* $0 \leq j \leq q_{n+1}$

$$\underset{I_n(x)}{Var} \log Df^j \leq \underset{[0,1]}{Var} \log Df = V \tag{3.3}$$

---

[1] If ψ has bounded variation and ϕ is Lipschitz with constant L_ϕ then $Var(\phi \circ \psi) \leq L_\phi \, Var \, \psi$

Proof. The chain rule applied to f^j gives:

$$\log Df^j = \sum_{i=0}^{j-1} \log Df \circ f^i. \tag{3.4}$$

The estimate (3.3) then follows from the first assertion of Lemma 3.7. $\square$

Proposition 3.9 has the following corollaries:

Corollary 3.10 *Let f, V, n and j be as in Proposition 3.9. Then*

$$e^{-V} \le \frac{|I_n(f^j(x))|}{|I_n(x)||Df^j(x)} \le e^V. \tag{3.5}$$

Proof. We first note that $I_n(f^j(x)) = f^j(I_n(x))$ and secondly that

$$|f^j(I_n(x))| \le |I_n(x)||Df^j(\xi)$$

for some $\xi \in I_n(x)$. Thus by (3.3) and Lemma 3.7 one has

$$\left| \log \frac{|I_n(f^j(x))|}{|I_n(x)||Df^j(x)} \right| \le \left| \log Df^j(\xi) - \log Df^j(x) \right| \le V. \quad \square$$

Corollary 3.11 *Let f, V, n and j be as in Proposition 3.9. Then*

$$|\log Df_n \circ f^j(x) - \log Df_n(x)| \le V \tag{3.6}$$

Proof. Since f_n and f^j commute we can write

$$\log Df_n \circ f^j + \log Df^j = \log Df^j \circ f_n + \log Df_n. \tag{3.7}$$

On the other hand, it follows from Proposition 3.9 that

$$|\log Df^j \circ f_n - \log Df^j| \le V. \tag{3.8}$$

$\square$

Corollary 3.12 (Denjoy's inequality) *Let f and V be as in Proposition 3.9. Then*

$$|\log Df_n|_{C^0} \le 2V \tag{3.9}$$

Proof. Since Df_n is a periodic function with mean value 1, there exists $x_0 \in \mathbb{R}$ such that $Df_n(x_0) = 1$, i.e. $\log Df_n(x_0) = 0$. Given $x \in \mathbb{R}$, take $0 \le j < q_{n+1}$ and $k \in \mathbb{Z}$ such that $x_1 = f^j T^k(x_0) \in J_n(x)$. Then we have:

$$|\log Df_n(x_1)| = |\log Df_n(x_1) - \log Df_n(x_0)| \le V,$$

and since[2] $x_1 \in I_n(x)$ or $x \in I_n(x_1)$, we also have

$$|\log Df_n(x) - \log Df_n(x_1)| \le V.$$

We can then conclude:

$$|\log Df_n(x)| \le |\log Df_n(x) - \log Df_n(x_1)| + |\log Df_n(x_1) - \log Df_n(x_0)|$$
$$\le 2V. \quad \square$$

Corollary 3.13 (Denjoy's Theorem) *All C^2 $\mathbb{Z}^2$-actions (3.2) with irrational rotation number are topologically conjugated to the standard action (2.16) by translations of $(-1, \alpha)$.*

Proof. This follows from Proposition 3.6 as Denjoy's inequality imply that the $(f_n)_{n \ge 0}$ are equicontinuous. $\square$

3.4   Denjoy's counter-examples

Denjoy has constructed examples of C^1 diffeomorphisms with no periodic orbits but not topologically conjugated to a rotation [De]. Indeed one can actually prove that

Theorem 3.14 *For every irrational α, there exists a diffeomorphism f of class $C^{2-\epsilon}$ for all $\epsilon > 0$, with rotation number equal to α which is not topologically conjugated to a rotation.*

We can find similar results in [DMVS,He1].

3.5   The Schwarzian derivative

To get a more precise control of the non-linearity of f_n we introduce the Schwarzian derivative $\mathcal{S}$. Let $F \in Diff_+^3(\mathbb{T})$ and f a lift of F, then the the Schwarzian derivative of f is defined by

$$\mathcal{S}f = D^2 \log Df - \frac{1}{2}(D \log Df)^2 = \frac{D^3 f}{Df} - \frac{3}{2}\left(\frac{D^2 f}{Df}\right)^2 \tag{3.10}$$

The following proposition collects some properties of $\mathcal{S}$.

Proposition 3.15 *Let $f \in Diff_+^3(\mathbb{R})$ then*

1. $\mathcal{S}(f \circ g) = (\mathcal{S}f) \circ g (Dg)^2 + \mathcal{S}g$;
2. *for $n \ge 1$ then $\mathcal{S}f^n = \sum_{i=0}^{n-1}(\mathcal{S}f) \circ f^i (Df^i)^2$;*
3. $|D \log Df|_{C^0} \le \sqrt{2}\sqrt{|\mathcal{S}f|_{C^0}}$.

---

[2] In fact if $x_1 \notin I_n(x)$ then $x_1 \in [f_n^{-1}(x), x]$, from which it follows that $x_1 \le x \le f_n(x_1)$, i.e. $x \in I_n(x_1)$.

Proof. 1. is immediate and 2. can be checked by induction using 1. To prove 3. choose x_0 such that $|D \log Df|_{C^0} = |D \log Df(x_0)|$, i.e. $D^2 \log Df(x_0) = 0$. Then

$$|D \log Df(x_0)| = \sqrt{2|Sf(x_0)|}$$

from which 3. follows. □

Corollary 3.16 *Let* $f \in Diff_+^3(\mathbb{R})$, $S = \|Sf\|_{C^0}$ *and* $M_n = \|f_n - id\|_{C^0}$. *For* $0 \leq j \leq q_{n+1}$ *one has*

$$|Sf^j(x)| \leq \frac{M_n e^{2V} S}{|I_n(x)|^2}$$

Proof. Using 2. of Proposition 3.15 we get

$$|Sf^j(x)| = \left| \sum_{i=0}^{j-1} (Sf) \circ f^i(x) \left(Df^i(x) \right)^2 \right|$$

$$\leq S \sum_{i=0}^{j-1} |Df^i(x)|^2$$

Using Corollary 3.10 we obtain:

$$|Sf^j(x)| \leq \frac{Se^{2V}}{|I_n(x)|^2} \sum_{i=0}^{j-1} |I_n(f^i(x))|^2.$$

Since $M_n = \max_x |I_n(x)|$, the trivial inequality $\sum a_i^2 \leq sup_i|a_i| \sum|a_i|$, and the disjointness of the intervals $I_n(f^i(x))$ lead to the desired conclusion. □

Corollary 3.17 *Let* $f \in Diff_+^3(\mathbb{R})$, $m_n = \min_x |I_n(x)|$ *and* $M_n = \max_x |I_n(x)|$. *For* $0 \leq j \leq q_{n+1}$ *one has*

$$|D \log Df^j|_{C^0} \leq \frac{c_1 M_n^{\frac{1}{2}}}{m_n}$$

with $c_1 = \sqrt{2S}e^V$.

Proof. Using 3. of Proposition 3.15 and Corollary 3.16 we have

$$|D \log Df^j|_{C^0} \leq \sqrt{2}\sqrt{|Sf^j|_{C^0}} \leq \sqrt{2}\frac{(M_n)^{\frac{1}{2}} e^V \sqrt{S}}{\min_x |I_n(x)|}. \square$$

Corollary 3.18 *Let* $f \in Diff_+^3(\mathbb{R})$, $m_n = \min_x |I_n(x)|$ *and* $M_n = \max_x |I_n(x)|$. *For* $0 \leq j \leq 2q_{n+1}$ *one has*

$$|D \log Df^j|_{C^0} \leq \frac{c_2 M_n^{\frac{1}{2}}}{m_n}$$

with $c_2 = \sqrt{2S}e^V (1 + e^{2V})$.

Proof. We only need to consider the case $q_{n+1} \leq j \leq 2q_{n+1}$, then $i = j - q_{n+1}$, varies between 0 and q_{n+1}. Applying Corollary 3.17 to $D \log Df^i \circ f_{n+1}$ the thesis follows. $\square$

Corollary 3.19 *Let* $f \in Diff^3_+(\mathbb{R})$, $M_n = \max_x |I_n(x)|$. *For* $m = n, n+1$ *one has*

$$|D \log Df_m(x)| \leq \frac{c_3 M_n^{\frac{1}{2}}}{|I_n(x)|}$$

where $c_3 = \sqrt{2S} e^{2V} (1 + e^{2V}) (2 + e^{2V})$.

Proof. Let $x_0 \in \mathbb{R}$ be such that $m_n = |I_n(x_0)|$, and let $x \in \mathbb{R}$. Choose $0 \leq j < q_{n+1}$, $k \in \mathbb{Z}$, $y \in J_n(x_0)$ such that $x = f^j T^k(y)$. One has

$$D \log D \left(f_m \circ f^j\right)(y) = D \log Df_m(x) Df^j(y) + D \log Df^j(y),$$

and therefore, by Corollaries 3.17, 3.18:

$$|D \log Df_m(x)| \leq \sqrt{2S} e^V (2 + e^{2V}) |Df^j(y)|^{-1}.$$

On the other hand, from Corollary 3.10, one has

$$|Df^j(y)|^{-1} \leq \frac{|I_n(y)|}{|I_n(x)|} e^V.$$

Finally, we have by Denjoy's inequality

$$|I_n(f_n(x_0))| \leq e^{2V} |I_n(x_0)|,$$
$$|I_n(f_n^{-1}(x_0))| \leq e^{2V} |I_n(x_0)|,$$

and therefore, as $I_n(y) \subset I_n(x_0) \cup I_n(f_n(x_0))$ or $I_n(y) \subset I_n(x_0) \cup I_n(f_n^{-1}(x_0))$:

$$|I_n(y)| \leq \left(1 + e^{2V}\right) |I_n(x_0)|$$

which proves the required inequality. $\square$

We can finally prove

Corollary 3.20 *Let* f *as in Proposition 3.15. Then* $|\log Df_n|_{C^0} \leq c_4 M_n^{\frac{1}{2}}$.

Proof. Let $x_0 \in \mathbb{T}$ such that $m_n = |I_n(x_0)|$ (so $\log Df_n(x_0) = 0$); from Lemma 3.7 we know that $J_n(x_0)$ cover the circle, so for all $x \in \mathbb{T}$ there exists $0 \leq j \leq q_{n+1}$ and $j \in J_n(x_0)$ such that $x = f^j(y)$. Letting $0 \leq j \leq 2q_{n+1}$ we can take $y \in I_n(x_0)$.

We use the fact that

$$\log Df^j(f_n(y)) - \log Df^j(y) = \log Df_n(x) - \log Df_n(y)$$

we write

$$\log Df_n(x) - \log Df_n(x_0) = \log Df^j(f_n(y)) - \log Df^j(y) \tag{3.11}$$
$$+ \log Df_n(y) - \log Df_n(x_0) \tag{3.12}$$

To estimate the right hand side of (3.11) we use

$$|\log Df^j(f_n(y)) - \log Df^j(y)| \le |\int_y^{f_n(y)} D\log Df^j(\xi)\, d\xi|$$

$$\le m_n(y)\,\|D\log Df^j\|_{C^0} \le \frac{|I_n(y)|}{|I_n(x_0)|}c_2 M_n^{\frac{1}{2}}$$

but $|f_n(y) - y| \le |f_n(y) - f_n(x_0)| + |x_0 - y| \le (e^V + 1)m_n(x_0)$.

For (3.12) we use something similar. And finally we obtain the result. $\square$

3.6   Partial renormalization

In this subsection we prove a first step of every renormalisation theory for circle diffeomorphisms. We prove that starting with some $f \in Diff_+^\omega(\mathbb{R})$, not necessarily near a translation, then for all $\epsilon > 0$ we can do an appropriate number of renormalisation steps in order to end with an analytic diffeomorphism f' which is ϵ close to a translation in the C^2 topology.

Proposition 3.21 *Let us consider the action* $\pi : \mathbb{Z}^2 \hookrightarrow Diff_+^\omega(\mathbb{R})$, *given by:*

$$\pi : e_{-1} \mapsto \pi(e_{-1}) = T$$
$$\pi : e_0 \mapsto \pi(e_0) = f$$

with $\rho(\pi) = [-1, \alpha]$ *and* $\alpha \in \mathbb{R} \setminus \mathbb{Q}$. *Let* $f_n = f^{q_n} \circ T^{p_n}$ *and* (f_n, f_{n+1}) *be a basis of the* $\mathbb{Z}^2$-action.

Let $\epsilon > 0$, *then for* n *large enough, there exists* $h_n \in Diff_+^\omega(\mathbb{R})$ *such that:*

$$h_n \circ f_n \circ h_n^{-1} = T$$
$$h_n \circ f_{n+1} \circ h_n^{-1} = F_{n+1}$$

with $F_{n+1} \in Diff_+^\omega(\mathbb{R})$, $\rho(F_{n+1}) = \alpha_{n+1}$ *and* $|D\log DF_{n+1}|_{C^0} < \epsilon$.

Proof. Let $x_0 \in \mathbb{T}$ be such that $m_n = |I_n(x_0)|$ and let A be the affine map such that $A(x_0) = 0$, $A(f_n(x_0)) = -1$. Set $\hat{f}_m = A \circ f_m \circ A^{-1}$ for $m = n, n+1$. By Corollaries 3.17 and 3.20, we have

$$\|\log D\hat{f}_m\|_{C^1} \le cM_n^{1/2}.$$

Moreover $\hat{f}_n(0) = -1$; therefore there exists $k_0 \in Diff_+^2([-1, 1])$ with $\|k_0 - \mathrm{id}\|_{C^2} < cM_n^{1/2}$ and $k_0 \circ \hat{f}_n \circ k_0^{-1} = T$ on $[0, 1]$. On the other hand, the quotient $\mathbb{R}/\hat{f}_n^{\mathbb{Z}}$ is a compact analytic 1-dimensional manifold, and is therefore C^ω-diffeomorphic to the circle: there exists $k_1 \in Diff_+^\omega([-1,1])$ with $k_1 \circ \hat{f}_n \circ k_1^{-1} = T$ on $[0,1]$.

The composition $k_0 \circ k_1^{-1}$ belongs to $Diff_+^2(\mathbb{T})$; taking $k \in Diff_+^\omega(\mathbb{T})$, C^2-close to $k_0 \circ k_1^{-1}$, and setting $k_2 = k \circ k_1$, we have $k_2 \in Diff_+^\omega([-1,1])$ and $\|k_2 - \mathrm{id}\|_{C^2} < cM_n^{1/2}$.

It is now sufficient to take $h_n = k_2 \circ A$: we have $h_n \circ f_n \circ h_n^{-1} = k_2 \hat{f}_n \circ k_2^{-1} = T$ and $F_{n+1} := h_n \circ f_{n+1} \circ h_n^{-1} = k_2 \circ \hat{f}_{n+1} \circ k_2^{-1}$, with

$$\| \log DF_{n+1} \|_{C^1} \leq cM_n^{1/2}. \qquad \square$$

4   Analytic case

4.1   A linearization criterion

Let $\pi : \mathbb{Z}^2 \to Diff_+^\omega(\mathbb{R})$ be an action with $\pi(e_{-1}) = T$, $\pi(e_0) = f$ as above. Let $\Delta > 0$ be such that f is holomorphic and univalent in B_Δ.

Proposition 4.1 *The action π is C^ω-linearizable if and only if the set*

$$L = \bigcap_{n \geq 0} f^{-n}(B_\Delta)$$

contains $\mathbb{R}$ in its interior.

Proof. The condition is obviously necessary. Assume that it is satisfied. Then the component U of $\mathrm{int}\,L$ which contains $\mathbb{R}$ is invariant under T and satisfies $f(U) \subset U$. It is simply connected by the maximum principle (applied to $\Im m\, f^n$); therefore there exists $\delta > 0$ and a real biholomorphism $h : B_\delta \to U$ commuting with T. Then $h^{-1} \circ f \circ h$ is real, commutes with T and sends univalently B_δ into itself. Therefore it is a translation. $\square$

4.2   Local Theorem 1.2: big strips

4.2.1   Scheme of the proof Let $\pi : \mathbb{Z}^2 \to Diff_+^\omega(\mathbb{R})$ be an action of $\mathbb{Z}^2$ with generators $\pi(e_{-1}) = T$, $\pi(e_0) = f$. We assume that $\rho(f) = \alpha$ is a Brjuno number and that f is holomorphic and univalent in a strip B_Δ with $\Delta > \frac{1}{2\pi}B(\alpha) + c$, where c is a conveniently large universal constant.

We set $F_0 = f$, $\Delta_0 = \Delta$, $\alpha_0 = \alpha$. We will denote by c_1, c_2, c_3 universal constants. We will construct, through a geometric construction explained in the next section, holomorphic maps H_n, F_n for $n > 0$ with the following properties:

- (i_n) $F_n \in Diff_+^\omega(\mathbb{R})$ commutes with T, has rotation number α_n, and is holomorphic and univalent in a strip B_{Δ_n} with

- (ii_{n+1}) $\Delta_{n+1} = \alpha_n^{-1} \left(\Delta_n - \frac{1}{2\pi} \log \alpha_n^{-1} - c_1 \right)$.
- (iii_{n+1}) The map H_{n+1} is holomorphic and univalent in the rectangle $R_n = \{ |\Re z| < 2, |\Im z| < \Delta_n - \frac{1}{2\pi} \log \alpha_n^{-1} - c_2 \}$. Moreover H_{n+1} restricted to $(-2, 2)$ is real and orientation reversing.
- (iv_{n+1}) For $z, z' \in R_n$ we have

$$|\alpha_n(\Im m\, H_{n+1}(z) - \Im m\, H_{n+1}(z')) + (\Im m\, z - \Im m\, z')| \leq c_3.$$

- (v_n) For $z \in R_n$ we have

$$\frac{9}{10}\alpha_n < \Re e(F_n(z) - z) < \frac{11}{10}\alpha_n$$

$$|\Im m(F_n(z) - z)| < \frac{1}{10}\alpha_n$$

- (vi_{n+1}) For $z \in R_n \cap T^{-1}R_n$

$$H_{n+1}(Tz) = F_{n+1}(H_{n+1}(z)) + a_{n+1}$$

- (vii_{n+1}) For $z \in R_n \cap F_n^{-1}(R_n)$

$$H_{n+1}(F_n(z)) = H_{n+1}(z) - 1.$$

Let us explain why these properties imply that the action generated by T, f is linearizable. We will apply the criterion of Section 4.1.

We can assume $c_3 \geq c_1 \geq 1$. Define, for $n \geq 0$

$$c_3(n) = c_3(1 + \alpha_n + \alpha_n\alpha_{n+1} + \alpha_n\alpha_{n+1}\alpha_{n+2} + \dots) < 4c_3.$$

Take $c = c_2 + 8c_3$. We will show that if

$$|Imz| < \Delta_n - \frac{1}{2\pi}B(\alpha_n) - [c_2 + 2c_3(n)]$$

then for all $m \geq 0$, we have

$$|\Im m\, F_n^m(z) - \Im m\, z| \leq c_3(n).$$

This allows to conclude by Proposition 4.1 (taking $n = 0$). Assume by contradiction that for some $n \geq 0$, some z and some $M > 0$, we have

$$|\Im m\, z| < \Delta_n - \frac{1}{2\pi}B(\alpha_n) - [c_2 + 2c_3(n)], \tag{4.1}$$

$$|\Im m\, F_n^m(z) - \Im m\, z| \leq c_3(n), \quad \text{for} \quad 0 \leq m < M \tag{4.2}$$

$$|\Im m\, F_n^M(z) - \Im m\, z| > c_3(n). \tag{4.3}$$

We may assume that $\Re z \in [0, 1)$. Observe that from (4.1), (4.2) we have

$$|\Im F_n^m(z)| < \Delta_n - \frac{1}{2\pi} B(\alpha_n) - [c_2 + 2c_3(n)]$$
$$< \Delta_n - \frac{1}{2\pi} \log \alpha_n^{-1} - (c_2 + 1)$$

for $0 \leq m < M$, hence by (v_n)

$$|\Im F_n^M(z)| < \Delta_n - \frac{1}{2\pi} \log \alpha_n^{-1} - c_2.$$

Thus, if we put $k_m = [\Re F_n^m(z)]$ for $0 \leq m \leq M$ and $w_m = F_n^m(z) - k_m$, we have $w_m \in R_n$ for all $0 \leq m \leq M$. We have also $k_0 = 0$ and $k_{m+1} - k_m \in \{0, 1, 2\}$ by (v_n). Also, $F_n(w_m) = w_{m+1} + (k_{m+1} - k_m)$. Set $u_m = H_{n+1}(w_m)$ for $0 \leq m \leq M$. By $(vi)_{n+1}$ and $(vii)_{n+1}$ we have

$$u_{m+1} = F_{n+1}^{k_{m+1}-k_m}(u_m) + (k_{m+1} - k_m)a_{n+1} - 1$$

and thus

$$u_m = F_{n+1}^{k_m}(u_0) + k_m a_{n+1} - m$$

for $0 \leq m \leq M$. On the other hand, by (4.3) and (iv_{n+1}), we have

$$|\alpha_n(\Im u_M - \Im u_0)| > c_3(n) - c_3$$

and therefore, as $c_3(n) = c_3 + \alpha_n c_3(n+1)$, we get

$$|\Im u_M - \Im u_0| > c_3(n+1). \tag{4.4}$$

We have also by (iv_{n+1}):

$$|\Im u_0| \leq \alpha_n^{-1}[|\Im z| + c_3]$$
$$< \alpha_n^{-1}[\Delta_n - \frac{1}{2\pi} B(\alpha_n) - (c_2 + 2c_3(n)) + c_3]$$

using (ii)

$$\leq \Delta_{n+1} + c_1\alpha_n^{-1} - \frac{1}{2\pi} B(\alpha_{n+1}) + c_3\alpha_n^{-1} - (c_2 + 2c_3(n))\alpha_n^{-1}$$
$$\leq \Delta_{n+1} - \frac{1}{2\pi} B(\alpha_{n+1}) - (c_2 + 2c_3(n+1)),$$

as $c_1 \leq c_3$. We conclude that u_0 for F_{n+1} and some integer $M' \leq k_M$ satisfy the same properties (4.1), (4.2), (4.3) of z for F_n and M. But we have, by (v_n)

$$k_M < \Re(F_n^M(z) - z) + 1 < \frac{11}{10}\alpha_n M + 1.$$

Iterating twice, we would have for F_{n+2} and some integer M'' a point t_0 with the same properties and

$$
\begin{aligned}
M'' &< \frac{11}{10}\alpha_{n+1}M' + 1 \\
&< \left(\frac{11}{10}\right)^2 \alpha_n \alpha_{n+1}M + \frac{11}{10}\alpha_{n+1} + 1 \\
&< \frac{1}{2}\left(\frac{11}{10}\right)^2 M + \frac{21}{10}.
\end{aligned}
$$

We conclude that for some $\bar{n}$ we would be able to find a point $\bar{z}$ satisfying (4.1), (4.2), (4.3) with $M \le 5$. But this is impossible because of $(v_{\bar{n}})$. As mentioned before this concludes the proof of Theorem 1.2.

4.2.2 The geometric construction The following result is due to Grötzsch [Ah]: Let $V \subset \mathbb{C}/\mathbb{Z}$ a homotopically non trivial annular open set, symmetric with respect to $\mathbb{R}/\mathbb{Z}$. Denote by $\mathrm{mod}(V)$ the modulus of V and by Δ_V the largest Δ such that $B_\Delta \subset V$. Then

$$
\Delta_V \ge \frac{1}{2}\mathrm{mod}(V) - \frac{1}{2\pi}\log 4.
$$

On the other hand, one obviously has

$$
\Delta_V \le \frac{1}{2}\mathrm{mod}(V).
$$

We write f in Section 4.2.1 as

$$
f(z) = z + \sum_{m \in \mathbb{Z}} \hat{f}(m)e^{2\pi i m z}.
$$

By Grötzsch's Theorem we have in $B_\Delta \cap \{\Im m\, z > 0\}$

$$
\Im m(f(z) - z) \ge -\frac{1}{2\pi}\log 4
$$

which gives for the Fourier coefficients universal estimates

$$
|\hat{f}(m)| \le e^{-2\pi|m|\Delta}c_4, \quad m \ne 0.
$$

We denote here and in the following by $c_4, c_5, \ldots$ universal constants. From this we easily get, for $0 \le \Im m\, z < \Delta - 1$

$$
|f(z) - z - \alpha_0| \le c_5 e^{-2\pi(\Delta - \Im m\, z)}, \tag{4.5}
$$

$$
|Df(z) - 1| \le c_6 e^{-2\pi(\Delta - \Im m\, z)}, \tag{4.6}
$$

$$
|D^2 f(z)| \le c_7 e^{-2\pi(\Delta - \Im m\, z)}. \tag{4.7}
$$

Thus there exists $c_8 > 0$ such that, if

$$0 \leq \Im m\, z \leq \Delta - \frac{1}{2\pi} \log \alpha_0^{-1} - c_8 = h$$

then

$$|f(z) - z - \alpha_0| \leq \frac{1}{10}\alpha_0, \tag{4.8}$$

$$|Df(z) - 1| \leq \frac{1}{10}\alpha_0, \tag{4.9}$$

$$|D^2 f(z)| \leq \frac{1}{10}\alpha_0. \tag{4.10}$$

Therefore, relation (v) in Section 4.2.1 will follow provided that $c_2 \geq c_8$.

Consider now the Jordan domain $U \subset B_\Delta$ whose boundary is the union of the vertical segment ℓ joining ih to $-ih$, the image $f(\ell)$ and the segments joining ih to $f(ih)$, $-ih$ to $f(-ih)$ (see Figure 1). By (4.8), (4.9) the curve we have described has no self-intersection. We glue the two vertical-like sides of U through the holomorphic map f; we thus obtain an abstract annular Riemann surface $\tilde{U}$. Complex conjugation leaves U invariant and induces an antiholomorphic automorphism of $\tilde{U}$. Define

$$\Delta' = \frac{1}{2}\mathrm{mod}(\tilde{U})$$

($\tilde{U}$ is obviously not biholomorphic to $\mathbb{C}^*$, thus $\Delta' < +\infty$). We now choose a biholomorphism $\tilde{H}_1$ from $\tilde{U}$ onto $B_{\Delta'}/\mathbb{Z}$, which is orientation reversing on the equators of these annular domains (sending the top boundary of $\tilde{U}$ onto the bottom one of $B_{\Delta'}/\mathbb{Z}$); we lift $\tilde{H}_1$ to an holomorphic map

$$H_1 : U \cup \ell \cup f(\ell) \hookrightarrow B_{\Delta'}$$

which satisfies

$$H_1(f(z)) = H_1(z) - 1, \tag{4.11}$$

for all $z \in \ell$.

To extend the domain of H_1 we will use the

Lemma 4.2 *Define*

$$R = \{|\Re e\, z| < 2, |\Im m\, z| < \Delta - \frac{1}{2\pi} \log \alpha^{-1} - c_2\}$$

with $c_2 = c_8 + 1$. If $z \in R$ and $\Re e\, z \geq 0$, there exists a unique integer $L \leq 0$ such that $f^L(z) \in U \cup \ell$ and

$$|\Im m\, f^j(z)| < \Delta - \frac{1}{2\pi} \log \alpha^{-1} - c_8$$

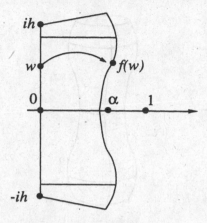

Fig. 1. The construction of the domain U

for $0 \geq j \geq L$. If $z \in R$ and $\Re e\, z < 0$, there exists a unique integer $L > 0$ such that $f^L(z) \in U \cup \ell$ and

$$|\Im m\, f^j(z)| < \Delta - \frac{1}{2\pi} \log \alpha^{-1} - c_8$$

for $0 \leq j \leq L$.

Proof. It is clear from (4.8). □

We now extend the domain of definition of H_1 to $U \cup R$ using the previous Lemma and the relation (4.11). By the Lemma, relation (4.11) will hold as soon as $z \in R \cap f^{-1}(R)$. Next we will define F_1 (from property (vi_1) in Section 4.2.1): set $V = H_1((U \cup \ell \cup f(\ell)) \cap R)$, $V^* = \cup_{n \in z} T^n(V)$, and denote by Δ^* the largest Δ_1 such that $B_{\Delta_1} \subset V^*$ (see Figure 2). For $z \in (U \cup \ell \cup f(\ell)) \cap R$, define

$$F_1(H_1(z)) = H_1(z-1) - a_1 \tag{4.12}$$

(which is possible as $z - 1 \in R$); if $z \in \ell \cap R$, we will have

$$
\begin{aligned}
F_1(H_1(z) - 1) &= F_1(H_1(f(z))) \\
&= H_1(f(z) - 1) - a_1 \\
&= H_1(f(z-1)) - a_1 \\
&= H_1(z-1) - 1 - a_1 \\
&= F_1(H_1(z)) - 1.
\end{aligned}
$$

This shows that F_1 extends from V to V^* as an holomorphic map, commuting with T. Also F_1 obviously restricts to a diffeomorphism of the real line. One

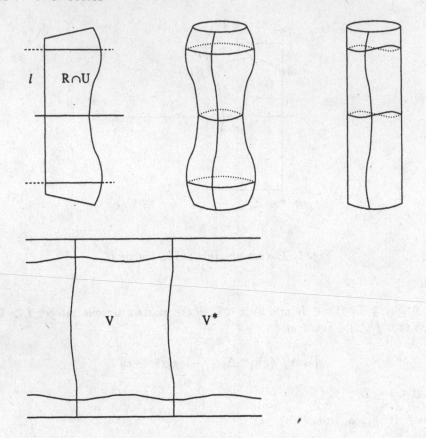

Fig. 2. Standard construction: glueing, uniformizing, developing on the plane

also easily checks that F_1 is univalent on V^*, and that relation (4.12) extends to all $z \in R \cap T^{-1}(R)$.

We now have constructed F_1 and H_1, and we have checked properties (i_1), (iii_1), (vi_1), (vii_1) of Section 4.2.1. We have also seen how (v) follows from the choice of R. There remains to check (ii_1) and (iv_1). Both follow from minorations of moduli of annular domains and Grötzsch's Theorem.

For $-h \leq y \leq h$, denote by σ_y the segment joining iy to $f(iy)$. For $0 < y < h$, let A_y (resp. A_y^+) be the Jordan domain whose boundary is formed of σ_0, σ_y, $[0, iy]$ and $f([0, iy])$ (resp. σ_h, σ_y, $[iy, ih]$ and $f([iy, ih])$). We glue through f the vertical sides of A_y, A_y^+ to obtain annular Riemann surfaces $\tilde{A}_y$, $\tilde{A}_y^+$.

In the Appendix we will show that

$$|\alpha \bmod \tilde{A}_y - y| \leq c_9, \tag{4.13}$$

$$|\alpha \bmod \tilde{A}_y^+ - (h - y)| \leq c_{10}. \tag{4.14}$$

We obtain (ii_1) (i.e. a bound from below for Δ^*) taking $y_0 = \Delta - \frac{1}{2\pi}\log\alpha^{-1} - c_2 - 1$. Then, $A_{y_0} \subset U \cap R$ and therefore

$$\frac{1}{2}\mathrm{mod}\,(V^*/\mathbb{Z}) \geq \alpha^{-1}(y_0 - c_9)$$

and (ii_1) follows by Grötzsch. To obtain (iv_1) consider $z \in R$ with $\Im m\, z > 0$. It is sufficient to consider $z' \in U \cup \ell$ (with $\Im m\, z' > 0$) because of (vii_1) and (v_0): the point $z' = f^L(z)$ given by the Lemma above satisfies

$$|\Im m\, z - \Im m\, z'| < 1.$$

On the other hand, such a z' lies on a unique segment σ_y with

$$|\Im m\, z' - y| \leq \frac{1}{10}.$$

From Grötzsch's Theorem, we now have a bound from below for $|\Im m\, H_1(z')|$ because

$$\mathrm{mod}\tilde{A}_y \geq \alpha^{-1}(y - c_9)$$

and a bound from above because

$$\mathrm{mod}\tilde{A}_y^+ \geq \alpha^{-1}(h - y - c_{10})$$

which together prove (iv_1).

This concludes the geometric part of the proof of the Theorem.

4.3   Local Theorem 1.3: small strips

The local Theorem 1.3 for small strips is an easy consequence of the local Theorem 1.2 for big strips.

Consider $f \in \mathit{Diff}_+^{\omega}(\mathbb{R})$, commuting with T, with rotation number $\alpha \in \mathcal{B}$, holomorphic in a strip B_Δ. We will now assume that

$$\varepsilon(f) = \sup_{B_\Delta}|f(z) - z - \alpha|$$

is small. Let $\eta \ll 1$ to be chosen later. We perform a construction similar to the one described in Section 4.2.2: consider the curve made of the segments joining $i(\Delta - \eta)$ to $-i(\Delta - \eta)$, $i(\Delta - \eta)$ to $f(i(\Delta - \eta))$ and $-i(\Delta - \eta)$ to $f(-i(\Delta - \eta))$ and the image $f([-i(\Delta - \eta), i(\Delta - \eta)])$; this is a Jordan curve if $\varepsilon(f)$ is sufficiently small, namely

$$\varepsilon(f) < \varepsilon_0(\Delta, \eta, \alpha)$$

(use Cauchy estimates to control $f([-i(\Delta - \eta), i(\Delta - \eta)])$).

Next we glue through f the vertical sides of the Jordan domain U bounded by this curve, to get an annular Riemann surface $\tilde{U}$. We also define

$$R = \{|\Re e\, z| < 2, |\Im m\, z| < \Delta - 2\eta\}.$$

As in Section 4.2.2, we can define a holomorphic map H_1 on R, uniformizing $\tilde{U}$, and a diffeomorphism F_1, commuting with T, and conjugated to Tf^{a_1} by H_1, holomorphic on a strip $B_{\Delta_1^*}$. The key fact is the following

Lemma 4.3 *Given $\Delta_1 < \alpha^{-1}\Delta$ and $\varepsilon_1 > 0$, we can choose $\eta \ll 1$ and ε_0 such that, if we start with $\varepsilon(f) < \varepsilon_0$, then we obtain F_1 holomorphic on B_{Δ_1} and*

$$\varepsilon(F_1) = \sup_{B_{\Delta_1}} |F_1(z) - z - \alpha_1| < \varepsilon_1.$$

The easy proof is omitted.

The local Theorem 1.3 now follows from the local Theorem 1.2 by iteration of the Lemma 4.3: for all $N > 0$, $\Delta_N < \beta_{N-1}^{-1}\Delta$, $\varepsilon_N > 0$, there exists $\bar{\varepsilon}_0 = \bar{\varepsilon}_0(\Delta_N, \varepsilon_N, \alpha, \Delta)$ such that, starting with $\varepsilon(f) < \bar{\varepsilon}_0$ we end up after N renormalization steps with a diffeomorphism $F_N \in Diff_+^\omega(\mathbb{R})$, commuting with T, holomorphic on B_{Δ_N}, conjugated to $T^{p_N} f^{q_N}$ and satisfying

$$\varepsilon(F_N) = \sup_{B_{\Delta_N}} |F_N(z) - z - \alpha_N| < \varepsilon_N.$$

Actually, by Cauchy estimates, we can as well assume that F_N is univalent on B_{Δ_N}. But then, if N is large enough, we can apply local Theorem 1.2 to F_N: indeed we have, if

$$\sum_{n \geq N} \beta_{n-1} \log \alpha_n^{-1} < \frac{\Delta}{2} \tag{4.15}$$

that

$$B(\alpha_N) = \beta_{N-1}^{-1} \sum_{n \geq N} \beta_{n-1} \log \alpha_n^{-1} < \frac{1}{2}\beta_{N-1}^{-1}\Delta.$$

Taking $\Delta_N = \frac{3}{4}\beta_{N-1}^{-1}\Delta$ and N large enough to have $\frac{1}{4}\beta_{N-1}^{-1}\Delta > c$ as well as (4.15), we conclude that F_N and then f are linearizable.

4.4   Global Theorem: complex Denjoy estimates

The estimates in the Proposition below will be used several times in the proof of the global conjugacy theorem. They are also of interest in order to control the dynamics in the non-linearizable case (see [PM]).

Let $f \in \mathit{Diff}^\omega_+(\mathbb{R})$ be a diffeomorphism commuting with T. We assume that f is holomorphic in a strip $\overline{B}_\Delta$, that Df does not vanish in this strip and we set

$$\tau = \sup_{\overline{B}_\Delta} |D \log Df(z)|$$

($\log Df$ being real on the real axis).

We assume that the rotation number α of f is irrational and denote by p_n/q_n its convergents. As in Section 3 we set $f_n = f^{q_n} \circ T^{p_n}$ and $M_n = \sup_{\mathbb{R}} |f_n(x) - x|$, and denote by $I_n(x)$ the interval with endpoints $x, f_n(x)$.

Proposition 4.4 *Assume that $\tau < \log 2$ and set $|w|_{\max} = \min\left(\frac{\Delta}{2M_n},\right.$ $\left.\frac{\log 2}{2\tau} - \frac{1}{2}\right)$. Let $w_0 \in C$ satisfy $|w_0| < |w|_{\max}$, let $x_0 \in \mathbb{R}$; define*

$$z_0 = x_0 + (f_n(x_0) - x_0)w_0.$$

Then the points $f^j(z_0)$, $0 \le j \le q_{n+1}$ belong to B_Δ; moreover, if we write

$$f^j(z_0) = f^j(x_0) + (f_n(f^j(x_0)) - f^j(x_0))w_j,$$

then we have, for $0 \le j \le q_{n+1}$:

$$\left|\log \frac{dw_j}{dw_0}\right| = \left|\log Df^j(z_0) - \log \frac{|I_n(f^j(x_0))|}{|I_n(x_0)|}\right| \le (2|w_0| + 1)\tau,$$

and

$$\left|\log \frac{w_j}{w_0}\right| \le (2|w_0| + 1)\tau < \log 2.$$

Proof. For every j, the second estimate is a consequence of the first. On the other hand, assume that the second inequality is valid for $0 \le l \le j$. Then we have, for $l < j$

$$|w_l| \le 2|w_0| < 2|w|_{max} < M_n^{-1}\Delta,$$

and thus $f^{j-1}(z_0) \in B_\Delta$ and $f^j(z_0)$ is defined. We have now, if $j \le q_{n+1}$:

$$\left|\log Df^j(z_0) - \log Df^j(x_0)\right| \le \sum_{l=0}^{j-1} \left|\log Df(f^l(z_0)) - \log Df(f^l(x_0))\right|$$

$$\le \sum_{l=0}^{j-1} |I_n(f^l(x_0))||w_l|\tau \le 2\tau|w_0|,$$

because $|w_l| \le 2|w_0|$ and $\sum_{l=0}^{q_{n+1}-1} |I_n(f^l(x_0))| \le 1$ as these intervals are disjoint. Similarly we get

$$\left|\log \frac{|I_n(f^j(x_0))|}{|I_n(x_0)|} - \log Df^j(x_0)\right| \le \tau,$$

which gives the first inequality (and hence the second) at step j.   □

As a consequence of the Proposition above and Proposition 3.21, we obtain the

Corollary 4.5 *Let* $\pi : \mathbb{Z}^2 \hookrightarrow Diff_+^\omega(\mathbb{R})$ *be an action with* $\pi(e_{-1}) = T$, $\pi(e_0) = f$, $\rho(\pi) = [-1, \alpha]$ *with irrational* α. *Write as before* f_n *for* $f^{q_n} \circ T^{p_n}$. *Then, for any* $\Delta_0 > 0$, *there exists* n *and a diffeomorphism* $k \in Diff_+^\omega(\mathbb{R})$ *such that*

$$k \circ f_n \circ k^{-1} = T$$
$$k \circ f_{n+1} \circ k^{-1} = F_{n+1}$$

where F_{n+1} *is holomorphic and univalent on* B_{Δ_0}, *with rotation number* α_{n+1}.

Proof. We may assume $\Delta_0 > 1$. Let $\eta \ll \Delta_0^{-1}$ be a small parameter to be determined later. By Proposition 3.21, we can find an analytic conjugacy for the action sending (for some $m = m(\eta, \Delta_0, f)$) f_m to T and f_{m+1} to a diffeomorphism F with

$$\sup_{\mathbb{R}} |D \log DF(x)| < \eta \Delta_0^{-1}.$$

Therefore, we can as well assume that $f = F$ satisfies this estimate. We then pick $\Delta > 0$ such that f is holomorphic and univalent on $\overline{B}_\Delta$ with

$$\tau := \sup_{\overline{B}_\Delta} |D \log DF(z)| < 2\eta \Delta_0^{-1}.$$

Then

$$\frac{\log 2}{2\tau} - \frac{1}{2} > \frac{\Delta_0}{8\eta}$$

provided that η is small enough. We choose n large enough to have also

$$\frac{\Delta}{2M_n} > \frac{\Delta_0}{8\eta} \quad \text{and} \quad M_n^{1/2} < \eta.$$

Then the estimates in the proposition hold for $|w_0| \leq (8\eta)^{-1}\Delta_0$. In particular, taking $\eta < 1/24$ and $w_0 \leq 3\Delta_0$, we will have, for $j \leq q_{n+1}$:

$$\left| \log \frac{dw_j}{dw_0} \right| \leq 14\eta, \tag{4.16}$$

$$\left| \log \frac{w_j}{w_0} \right| \leq 14\eta. \tag{4.17}$$

and also, by Corollary 3.20

$$\left| \log \frac{|I_n(f_n(x))|}{|I_n(x)|} \right| < c_4\eta, \tag{4.18}$$

$$\left| \log \frac{|I_n(f_{n+1}(x))|}{|I_n(x)|} \right| < c_4\eta. \tag{4.19}$$

We now perform the same geometric construction as in Section 4.2 or 4.3: set $h = 2\Delta_0 |I_n(0)|$; for $\eta \ll \Delta_0^{-1}$, consider the Jordan domain U whose "vertical" sides are the segment $[-ih, ih]$ and its image by f_n, and horizontal sides are $[-ih, f_n(-ih)]$, $[ih, f_n(ih)]$; glue the vertical sides of U through f; uniformize the resulting annular Riemann surface to get an holomorphic map $k : U \to \mathcal{C}$ which is real and orientation-reversing on $U \cap \mathbb{R}$, extends continuously to the vertical sides of U, and satisfies

$$k(f_n(it)) = k(it) - 1 \qquad (4.20)$$

for $|t| < h$. Next, defining the rectangle

$$R = \left\{ |\Re e\, z| \leq 3|I_n(0)|, |\Im m\, z| \leq \frac{3}{4}h \right\},$$

we use (4.16) and (4.18) (with $\eta \ll \Delta_0^{-1}$) to extend through relation (4.20) the domain of k to $U \cup R$. Setting

$$U_1 = \left\{ z \in U, |\Im m\, z| < \frac{2}{3}h \right\},$$

we deduce from (4.16), (4.18) that

$$f_{n+1}(U_1) \subset R$$

and this allows to define a holomorphic map F_{n+1} on $k(U_1)$ by

$$F_{n+1}(k(z)) = k(f_{n+1}(z)), z \in U_1.$$

We will be finished if the strip B_{Δ_0} is contained in the union $\cup_{m \in \mathbb{Z}} T^m(k(U_1))$. But this will be clearly true if η is small enough. $\square$

4.5   Global Theorem: proof of linearization

4.5.1   The geometric construction with large strip but small rotation number

Let $\pi : \mathbb{Z}^2 \hookrightarrow Diff_+^\omega(\mathbb{R})$ be an analytic action with $\pi(e_{-1}) = T$, $\pi(e_0) = f$, $\rho(\pi) = [-1, \alpha]$, $\alpha \in (0, 1)$. We assume that f is holomorphic and univalent in a strip B_Δ, with $\Delta \geq \Delta_{min}$, Δ_{min} a universal constant to be determined later. We also assume that α is so small that

$$\Delta \leq \frac{1}{2\pi} \log \alpha^{-1} + c_0, \qquad (4.21)$$

with c_0 another universal constant determined as follows: if on the opposite $\Delta > \frac{1}{2\pi} \log \alpha^{-1} + c_0$, we are able to perform the construction of Section 4.2.2 with the estimates indicated in Section 4.2.1.

From relations (4.5), (4.6) and (4.7) we get, for $|\Im m\, z| \leq 1$:

$$|f(z) - z - \alpha| \leq c_5' e^{-2\pi\Delta}, \tag{4.22}$$

$$|Df(z) - 1| \leq c_6' e^{-2\pi\Delta}, \tag{4.23}$$

$$|D\log Df(z)| \leq c_7' e^{-2\pi\Delta}, \tag{4.24}$$

and then, joining (4.21) with (4.22)

$$|f(z) - z| \leq c_{11} e^{-2\pi\Delta}. \tag{4.25}$$

We apply Proposition 4.4 to f in the strip B_1 with $n = 0$. We have

$$|w|_{\max} = \min\left(\frac{1}{2M_0}, \frac{\log 2}{2\tau} - \frac{1}{2}\right) \geq c_{12} e^{2\pi\Delta}.$$

Let $h = \theta e^{2\pi\Delta}|f(0)|$, with θ a universal constant conveniently small, in particular $\theta < \frac{1}{2}c_{12}$. If we have also $c_6'\theta \ll 1$, then we can define a Jordan domain U whose boundary is formed of $[-ih, ih]$, its image under f, and the segments $[-ih, f(-ih)]$, $[ih, f(ih)]$. We glue the vertical sides of U through f, and uniformize the resulting Riemann surface to get a map $H : U \to \mathbb{C}$, real and orientation-reversing on $U \cap \mathbb{R}$, which extends continuously to the vertical sides of U and satisfies

$$H(f(z)) = H(z) - 1, z \in \ell. \tag{4.26}$$

For small θ, we have from (4.23)

$$|f(z) - z - f(0)| \leq \frac{1}{10}|f(0)| \tag{4.27}$$

if $|\Re e\, z| \leq 3|f(0)|$, $|\Im m\, z| \leq h$. This allows to extend the domain of H, through (4.26), to

$$R = \left\{|\Re e\, z| \leq 3|f(0)|, |\Im m\, z| \leq \frac{3}{4}h\right\}.$$

Now, Proposition 4.4 shows that (provided θ is small) if $z \in \overline{U}$ and $|\Im m\, z| \leq \frac{1}{2}h$, then $Tf^{a_1}(z) \in R$. Therefore we can define F on $V = H(\overline{U} \cap B_{h/2})$ by

$$F(H(z)) = H(Tf^{a_1}(z)), z \in \overline{U} \cap B_{h/2}. \tag{4.28}$$

Then F will extend to an holomorphic and univalent map on $\cup_{m\in\mathbb{Z}}T^m(V)$, commuting with T. It remains to estimate the witdth of the largest strip B_Δ contained in $\cup_{m\in\mathbb{Z}}T^m(V)$. By the Grotzsch's Theorem mentioned earlier, this amounts to obtain a bound from below for the modulus of the annular Riemann surface $\tilde{V}$ obtained by glueing through f the vertical sides of V. In the appendix, we will show that

$$\mathrm{mod}\tilde{V} > c_{13}\theta e^{2\pi\Delta}$$

which gives finally (with now a fixed value of θ)

$$\Delta^* \geq c_{14} e^{2\pi\Delta}. \tag{4.29}$$

4.5.2   Conclusion of the proof of the global theorem The proof of the global theorem is now easy from the above estimates and the local theorem. Recall from Section 2.5 the definition (2.11) of the functions r_α ($\alpha \in (0,1)$) involved in the statement of conditon $\mathcal{H}$. On the other side, we define, in connection with 4.2.1 (ii) and (4.29), for $\alpha \in (0,1)$, $\Delta \in \mathbb{R}$

$$r_\alpha^*(\Delta) = \begin{cases} \alpha^{-1}\left(\Delta - \frac{1}{2\pi}\log\alpha^{-1} - c_1\right) & \text{if } \Delta \geq \frac{1}{2\pi}\log\alpha^{-1} + c_0, \\ c_{14}e^{2\pi\Delta} & \text{if } \Delta < \frac{1}{2\pi}\log\alpha^{-1} + c_0, \end{cases} \quad (4.30)$$

where c_0 has been chosen such that $c_0 > c_1 + 1$.

Lemma 4.6 *For any $c > 0$, there exists $\overline{\Delta}_0^*$ such that, for any $\alpha \in (0,1)$, any $\Delta_0^* \geq \overline{\Delta}_0^*$, any $k \geq 0$, one has*

$$r_{\alpha_{k-1}}^* \circ \cdots \circ r_{\alpha_0}^*(\Delta_0^*) \geq c + \frac{1}{2\pi}r_{\alpha_{k-1}} \circ \cdots \circ r_{\alpha_0}(0).$$

Proof. Left as an exercise to the reader. $\square$

Let now $\alpha \in \mathcal{H}$ and $\pi : \mathbb{Z}^2 \to Diff_+^\omega(\mathbb{R})$ be an action with $\pi(e_{-1}) = T$, $\pi(e_0) = f$, $\rho(\pi) = [-1, \alpha]$. With the constant c involved in the statement of the local theorem (large strips), we apply the Lemma above which gives a constant $\overline{\Delta}_0^*$. We then make use of Corollary 4.5: we find N and $k \in Diff_+^\omega(\mathbb{R})$ such that

$$k \circ f_{N-1} \circ k^{-1} = T,$$
$$k \circ f_N \circ k^{-1} = F,$$

with F holomorphic and univalent in $B_{\overline{\Delta}_0^*}$ and $\rho(F) = \alpha_N$. For $k > 0$, we now set

$$\Delta_k^* = r_{\alpha_{N+k-1}}^* \circ \cdots \circ r_{\alpha_N}^*(\overline{\Delta}_0^*),$$
$$x_k = r_{\alpha_{N+k-1}} \circ \cdots \circ r_{\alpha_N}(0).$$

By the very definition of condition $\mathcal{H}$, there exists K such that

$$x_K \geq B(\alpha_{N+K})$$

which implies, by the Lemma above

$$\Delta_K^* \geq c + \frac{1}{2\pi}B(\alpha_{N+K}).$$

On the other side, starting from F, the geometric constructions of Sections 4.2.2 or 4.5.1 allow to obtain a sequence of diffeomorphisms F_k, $0 \leq k \leq K$ such that

- the action generated by T, F_k is analytically conjugated to that generated by T, F;
- F_k is holomorphic and univalent in $B_{\Delta_k^*}$, with rotation number α_{N+k}.

But Δ_K^* is large enough to apply the local theorem: the action generated by T, F_K (and thus also the initial action) is analytically linearizable.

4.6 Global Theorem: Construction of nonlinearizable diffeomorphisms

4.6.1 Introduction In this last section, we will construct, for any $\alpha \notin \mathcal{H}$, analytic actions with rotation number $[-1, \alpha]$ which are not linearizable. Actually we get even a slightly more precise result, relating the way in which condition $\mathcal{H}$ fails to the width of the strips of univalence for the examples.

The construction of nonlinearizable actions is based on a geometric construction analogous to, but more sophisticated than, those which were made in Section 4.2.2. It is explained in Section 4.6.4. In the next section, we will state the result of this geometric construction; in Section 4.6.3, we explain how to derive the construction of analytic nonlinearizable actions from the basic step 4.6.2.

4.6.2 Result of the geometric construction In this section, we are given an analytic action $\pi : \mathbb{Z}^2 \to Diff_+^\omega(\mathbb{R})$ with $\pi(e_{-1}) = T$, $\pi(e_0) = f$ and $\rho(\pi) = [-1, \alpha]$, with $\alpha > 1$. We assume that f is holomorphic and univalent in a strip B_Δ.

Proposition 4.7 *There exist universal constants $c_{15}, c_{16}, \dots, c_{20}$ such that, if $\Delta > c_{15}$ we are able to construct an analytic diffeomorphism F, commuting with T, with rotation number α^{-1} and the following properties:*

(i)   *F is holomorphic and univalent in a strip $B_{\Delta'}$ with*

$$\Delta' = \begin{cases} \alpha^{-1}\Delta + \frac{1}{2\pi}\log\alpha - c_{17} & \text{if } \alpha < c_{16}\Delta \\ \frac{1}{2\pi}\log\Delta - c_{18} & \text{if } \alpha \geq c_{16}\Delta. \end{cases}$$

(ii)   *If $\alpha \geq c_{16}\Delta$, F has a fixed point p with $0 < \Im m\, p < c_{19} < \Delta'$.*

(iii)   *Assume that, for some $P, Q > 0$, f has a periodic orbit $(mod\,\mathbb{Z})$ $\mathcal{O}$ with*

$$f^Q(z) - P = z, z \in \mathcal{O},$$
$$0 < \max_{\mathcal{O}} \Im m\, z < 2c_{19};$$

then F has a periodic orbit $\mathcal{O}'$ with

$$f^P(z') - Q = z', z' \in \mathcal{O}',$$
$$0 < \max_{\mathcal{O}'} \Im m\, z' < \lambda \max_{\mathcal{O}} \Im m\, z,$$

where

$$\lambda = \begin{cases} \frac{4}{3}\alpha^{-1} & \text{if } \alpha < c_{16}\Delta \\ c_{20}\Delta^{-1} & \text{if } \alpha \geq c_{16}\Delta. \end{cases}$$

We defer the proof of this proposition to Section 4.6.4.

4.6.3   Construction of nonlinearizable actions Before we construct, using Proposition 4.7 as a building block, nonlinearizable analytic diffeomorphisms, we need a few preliminary arithmetical remarks.

For $\alpha > 1$, $\Delta > 0$, we define, in connection to Section 4.6.2:

$$R_\alpha^*(\Delta) = \begin{cases} \alpha^{-1}\Delta + \frac{1}{2\pi}\log\alpha - c_{17} & \text{if} \quad \Delta > c_{16}^{-1}\alpha, \\ \frac{1}{2\pi}\log\Delta - c_{18} & \text{if} \quad \Delta \le c_{16}^{-1}\alpha. \end{cases} \tag{4.31}$$

This is closely connected to the function $(r_{\alpha^{-1}})^{-1}$:

$$(r_{\alpha^{-1}})^{-1}(x) = \begin{cases} \alpha^{-1}x + \log x + 1 & \text{if} \quad x \ge \alpha, \\ \log x & \text{if} \quad x \le \alpha. \end{cases} \tag{4.32}$$

The function R_α^* appears in the third of the following elementary lemmas.

Lemma 4.8 *Let α be irrational and $x_0 \ge 0$. Assume that for all $k \ge 0$ we have*

$$x_k := r_{\alpha_{k-1}} \circ \cdots \circ r_{\alpha_0}(x_0) \ge \log\alpha_k^{-1}.$$

Then $\alpha \in \mathcal{B}$ and $x_k \ge B(\alpha_k) - 4$ for all $k \ge 0$.

Lemma 4.9 *Let α be irrational and $x^* \ge 0$. Then $\alpha \notin \mathcal{H}$ if and only if there exists m and an infinite set $I = I(m, x^*, \alpha)$ such that, for all $k \in I$*

$$r_{\alpha_{m+k-1}} \circ \cdots \circ r_{\alpha_m}(x^*) < \log\alpha_{m+k}^{-1}.$$

Lemma 4.10 *There exists a universal constant $c_{21} > 0$ with the following property. Let α be irrational, $x_0 > 2\pi c_{21}$, $N > 0$; define*

$$x_k := r_{\alpha_{k-1}} \circ \cdots \circ r_{\alpha_0}(x_0), \quad 0 \le k \le N,$$

choose $\Delta_N \ge \frac{1}{2\pi}x_N + c_{21} - \frac{1}{2\pi}x_0$ and define

$$\Delta_k := R_{\alpha_k^{-1}}^* \circ \cdots \circ R_{\alpha_{N-1}^{-1}}^*(\Delta_N), \quad 0 \le k \le N.$$

Then, for all $0 \le k \le N$, we have

$$\Delta_k \ge \frac{1}{2\pi}x_k - c_{21} - \max\left(0, \frac{1}{2\pi}x_N - \Delta_N\right).$$

All the three lemmas above are elementary computations that are left to the reader.

We now select some irrational number α which does not belong to $\mathcal{H}$, and some width $\Delta^* \ge 1$. We set

$$c_{22} = c_{18} + c_{21} + \frac{1}{2\pi}\log^+ c_{16},$$

$$x^* = 2\pi(\Delta^* + c_{22}),$$

and we fix some integer m given by Lemma 4.9 such that the set $I = I(m, x^*, \alpha)$ in this lemma is infinite. We will construct nonlinearizable analytic diffeomorphisms with rotation number α_m which are holomorphic and univalent in B_{Δ^*}.

Let $k \in I$. We define diffeomorphisms $f_l^{(k)}$, $0 < l \leq k + 1$, with rotation number α_{m+l-1}^{-1} and diffeomorphisms $F_l^{(k)}$, $0 \leq l \leq k$, with rotation number α_{m+l} as follows:

- $f_{k+1}^{(k)} = R_{\alpha_{m+k}^{-1}}$;
- $F_l^{(k)}$ is obtained from $f_{l+1}^{(k)}$ by Proposition 4.7;
- $f_l^{(k)} = T^{-a_{m+l}} \circ F_l^{(k)}$.

We choose to consider $f_{k+1}^{(k)} = R_{\alpha_{m+k}^{-1}}$ on the strip $B_{\Delta_{k+1}^{(k)}}$ with $\Delta_{k+1}^{(k)} = c_{16}^{-1} \alpha_{m+k}^{-1}$. It then follows from Proposition 4.7 that $F_l^{(k)}$ and $f_l^{(k)}$ are holomorphic and univalent on a strip $B_{\Delta_l^{(k)}}$ with

$$\Delta_l^{(k)} = R_{\alpha_{m+l}^{-1}}^* \circ \cdots \circ R_{\alpha_{m+k}^{-1}}^* (\Delta_{k+1}^{(k)}).$$

Let

$$x_l = r_{\alpha_{m+l-1}} \circ \cdots \circ r_{\alpha_m}(x^*).$$

Observe that we have, as $k \in I$:

$$\Delta_k^{(k)} = \frac{1}{2\pi} \log \alpha_k^{-1} - c_{22} + c_{21} \geq \frac{1}{2\pi} x_k + c_{21} - c_{22}$$

and therefore, by Lemma 4.10

$$\Delta_l^{(k)} \geq \frac{1}{2\pi} x_l - c_{22}.$$

In particular, taking $l = 0$, we have

$$\Delta_0^{(k)} \geq \Delta^*.$$

On the other hand, by (ii) in Proposition 4.7, for $k' \in I$, $k' \leq k$, $F_{k'}^{(k)}$ has a fixed point $p_{k'}^{(k)}$ such that

$$0 < \Im m \, p_{k'}^{(k)} < c_{19}.$$

We take $\Delta^* > 2c_{20}$ and iterate (iii) in Proposition 4.7; we obtain that $F_0^{(k)}$ has a periodic orbit $\mathcal{O}_{k'}^{(k)}$ (mod $\mathbb{Z}$) of fixed period $Q_{k'}$ with

$$0 < \max_{\mathcal{O}_{k'}^{(k)}} \Im m \, z < 2 \left(\frac{9}{10} \right)^{k'} c_{19}.$$

Finally, we extract from $(F_0^{(k)})_{k \in I}$ a subsequence converging to $F_0^{(\infty)}$: this is an analytic diffeomorphism, holomorphic and univalent in B_{Δ_*}, with rotation number α_m, which has, for each $k \in I$, a periodic orbit $\mathcal{O}_k^{(\infty)}$ (mod $\mathbb{Z}$) of period Q_k satisfying

$$0 < \max_{\mathcal{O}_k^{(\infty)}} \Im m \, z \le 2 \left(\frac{9}{10} \right)^k c_{19}.$$

Therefore $F_0^{(\infty)}$ is not linearizable.

Remark 4.11 To obtain $p_{k'}^{(k)}$ from (ii) in Proposition 4.7, we need $\Delta_{k'+1}^{(k)} \le c_{16}^{-1} \alpha_{m+k'}^{-1}$; this is obtained by definition for $k' = k$; when $k' < k$ we can force this inequality by taking $\Delta_{k'+1}^{(k)}$ smaller if necessary.

4.6.4 Proof of the Proposition 4.7 Let f, α, Δ be as in Proposition 4.7, with $\Delta > c_{15}$, c_{15} large enough.

By the estimates at the beginning of Section 4.2.2, there exists c_{23} such that, for $0 \le \Im m \, z \le \Delta - c_{23}$, we have

$$|f(z) - z - \alpha| \le \frac{1}{10} \exp[-2\pi(\Delta - c_{23} - \Im m \, z)],$$

$$|Df(z) - 1| \le \frac{1}{10} \exp[-2\pi(\Delta - c_{23} - \Im m \, z)],$$

$$|D^2 f(z)| \le \frac{1}{10} \exp[-2\pi(\Delta - c_{23} - \Im m \, z)].$$

On the other hand, we choose a $\mathbb{Z}$-periodic function ψ in $\{\Im m \, z \ge \Delta - c_{23} + 1\}$ which satisfies there

$$|\psi(z)| \le \frac{1}{10} \exp[-2\pi(\Im m \, z - \Delta + c_{23} - 1)],$$

$$|D\psi(z)| \le \frac{1}{10} \exp[-2\pi(\Im m \, z - \Delta + c_{23} - 1)],$$

$$|D^2 \psi(z)| \le \frac{1}{10} \exp[-2\pi(\Im m \, z - \Delta + c_{23} - 1)].$$

We also choose $\alpha' \in \mathbb{C}$, with $|\alpha - \alpha'| \le \frac{1}{10}$ and set

$$g(z) = z + \alpha' + \psi(z), \quad \Im m \, z \ge \Delta - c_{23} + 1.$$

We define also

$$\bar{g}(z) = \overline{g(\bar{z})}, \quad \Im m \, z \le -(\Delta - c_{23} + 1).$$

The most obvious and easiest choices are $\psi \equiv 0$, $\alpha' = \alpha$, but other choices are allowed and may give interesting properties.

We define a vertical strip U as follows. The left boundary is the imaginary axis, that we view as the union of the segments $[-i(\Delta - c_{23}), i(\Delta - c_{23})] = \ell$, $[i(\Delta - c_{23}), i(\Delta - c_{23} + 1)] = \ell^*$, $[i(\Delta - c_{23} + 1), +i\infty) = \ell^+$ and the symmetric $\bar{\ell}^*$ and $\bar{\ell}^+$.

The right boundary is the union of the curves $f(\ell)$, $g(\ell^+)$, $g(\bar{\ell}^+)$, and the two (symmetric) straight segments ℓ_1^*, $\bar{\ell}_1^*$ which connect $f(i(\Delta - c_{23}))$ to $g(i(\Delta - c_{23} + 1))$ and $f(-i(\Delta - c_{23}))$ to $\bar{g}(-i(\Delta - c_{23} + 1))$.

The estimates on f, ψ above guarantee that the domain U is well defined. We will also have to consider the square

$$S = \{0 \leq \Re e\, z \leq 1, \Delta - c_{23} \leq \Im m\, z \leq \Delta - c_{23} + 1\}$$

and its symmetric $\overline{S}$.

Next, we glue together some parts of the boundary of U: we glue ℓ to $f(\ell)$ through f, ℓ^+ to $g(\ell^+)$ through g and $\bar{\ell}^+$ to $\bar{g}(\bar{\ell}^+)$ through $\bar{g}$. We obtain a Riemann surface $\tilde{U}$ which is a twice-punctured annulus, the two punctures corresponding to $+i\infty$ and $-i\infty$ and the two components of $\partial\tilde{U}$ to $\ell^* \cup \ell_1^*$, $\bar{\ell}^* \cup \bar{\ell}_1^*$ (see Figures 3 and 4).

The fact that $\pm i\infty$ give rise to punctures (and not holes) is checked via moduli estimates from the appendix.

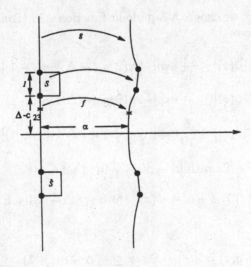

Fig. 3.

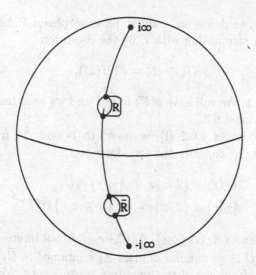

Fig. 4.

Denote by $\tilde{U}_1$ the trace in $\tilde{U}$ of $\overline{U} \setminus (S \cup \overline{S})$. With $z \in \overline{U} \setminus (S \cup \overline{S})$, the formulas

$$z \mapsto z - 1 \quad \text{if} \quad \Re e\, z \geq 1,$$
$$z \mapsto f(z-1) \quad \text{if} \quad 0 \leq \Re e\, z \leq 1, |\Im m\, z| < \Delta - c_{23},$$
$$z \mapsto g(z-1) \quad \text{if} \quad 0 \leq \Re e\, z \leq 1, \Im m\, z > \Delta - c_{23} + 1,$$
$$z \mapsto \bar{g}(z-1) \quad \text{if} \quad 0 \leq \Re e\, z \leq 1, \Im m\, z < -(\Delta - c_{23} + 1),$$

induce after glueing a map $\tilde{F} : \tilde{U}_1 \to \tilde{U}$ which is holomorphic and univalent. Moreover, $\tilde{F}$ extends holomorphically at the punctures and leave these punctures fixed. We denote by $\tilde{\tilde{U}}$ (resp. $\tilde{\tilde{U}}_1$) the Riemann surface $\tilde{U}$ (resp. $\tilde{U}_1$) with the punctures filled in. Let

$$2\tilde{\Delta} = \text{mod}\tilde{U},$$
$$2\tilde{\Delta}_1 = \text{mod}\tilde{U}_1,$$

and let $\tilde{H} : \tilde{U} \to B_{\tilde{\Delta}/\mathbf{Z}}$ be a biholomorphism which is orientation-reversing on the equators. Let $H : \overline{U} \setminus (\ell^* \cup \ell_1^* \cup \bar{\ell}^* \cup \bar{\ell}_1^*) \to \mathbb{C}$ a lift of $\tilde{H}$. Define

$$V = H(\overline{U} \setminus (S \cup \overline{S})),$$
$$V^* = \cup_{n \in \mathbf{Z}} T^n(V).$$

We have $V^*/\mathbf{Z} = \tilde{H}(\tilde{U}_1)$. By Grotzsch's theorem, setting $\Delta' = \tilde{\Delta}_1 - \frac{1}{2\pi} \log 4$, we have

$$B_{\Delta'} \subset V^*.$$

On the other side, we define an analytic diffeomorphism F, holomorphic and univalent on $B_{\Delta'}$, commuting with T by the condition

$$H(z-1) = F(H(z)),$$

for $z \in U$, $\Re e\, z > 1$. We will have $\rho(F) = \alpha^{-1}$ and we need to check the other properties of Section 4.6.2.

To get the estimates 4.6.2 (i), we need to bound $\tilde{\Delta}_1$ from below. Let $\omega = i(\Delta - c_{23} + 1/2)$; consider the sets, for $r > 2$

$$A_L(r) = \{2 < |z - \omega| < r\} \cap \overline{U},$$
$$A_R(r) = \{2 < |z - (\omega + \alpha)| < r\} \cap \overline{U}.$$

If $\Delta - c_{23} < \frac{\alpha}{2}$, the sets $A_L(\Delta - c_{23})$, $A_R(\Delta - c_{23})$ do not intersect and the trace of their union in $\tilde{U}$ is an annular domain A_1 contained in $\tilde{U}_1$ and separating the equator of $\tilde{U}_1$ from one of the boundary components. We therefore have

$$\tilde{\Delta}_1 \geq \mathrm{mod}\, A_1$$

and we will show in the appendix that

$$\mathrm{mod}\, A_1 \geq \frac{1}{2\pi} \log \Delta - c'_{18}$$

provided that Δ is large enough ($\Delta > c_{15}$).

If $\Delta - c_{23} \geq \frac{\alpha}{2}$, the sets $A_L(\alpha/2)$, $A_R(\alpha/2)$ do not intersect and the trace of their union in $\tilde{U}$ is an annular domain A_2 contained in $\tilde{U}_1$ and separating the equator of $\tilde{U}_1$ from one of the boundary components. We will show in the appendix that we have

$$\mathrm{mod}\, A_2 \geq \frac{1}{2\pi} \log \alpha - c'_{17}.$$

When moreover $\Delta - c_{23} \geq \alpha$, we also consider the annular domain $A'_2 \subset \tilde{U}_1$ whose trace on U is the domain between the segments $[0, f(0)]$ and $[\omega_1, f(\omega_1)]$, with $\omega_1 = i\left(\Delta - c_{23} - \frac{\alpha}{2}\right)$. This annular domain is disjoint from A_2 and separates the equator of $\tilde{U}_1$ from the same boundary component (associated to S) than A_2. In the appendix, we will show that

$$\mathrm{mod}\, A'_2 \geq \frac{\Delta}{\alpha} - c''_{17}.$$

We will therefore have

$$\tilde{\Delta}_1 \geq \frac{1}{2\pi} \log \alpha - c'_{17},$$

if $\Delta - c_{23} \geq \alpha/2$, and even

$$\tilde{\Delta}_1 \geq \frac{1}{2\pi} \log \alpha + \alpha^{-1}\Delta - c'_{17} - c''_{17},$$

when $\Delta - c_{23} \geq \alpha$.

The three estimates for $\tilde{\Delta}_1$ give (i) in 4.6.2. We turn now to (ii). The two symmetric fixed points we are interested in (when $c_{16}\Delta \leq \alpha$) are $H(\pm i\infty) = p^{\mp}$. We define the following annular domain in $\tilde{U}$: in $\tilde{U}$ we take out $\{p^-, p^+\}$, the trace of the line $\{\Re e\, z = \alpha/2\}$ and the trace of the segment $[-i\omega, i\omega]$; this is an annular domain A_3 which separates $\{p^+, p^-\}$ from the boundary components of $\tilde{U}$. In the appendix we will show that

$$\mathrm{mod} A_3 \geq \frac{1}{2\pi} \log(\alpha\Delta^{-1}) - c'_{19} \geq c''_{19} > 0.$$

Considering the uniformization $\tilde{H}$ of $\tilde{U}$ (filled-in at $\pm i\infty$), we obtain (ii) from standard Teichmüller estimates (see [Ah]). Finally, we consider property (iii) of 4.6.2. We will assume $c_{15} > 2c_{19} + c_{23} + 1$. Let $\mathcal{O}$ be a periodic orbit as in 4.6.2 (iii); let $z_0 \in \mathcal{O}$ with $0 \leq \Re e\, z_0 < 1$. We define inductively z_i, $i \geq 0$, by

$$z_{i+1} = z_i - 1 \quad \text{if} \quad \Re e\, z_i \geq 0,$$
$$z_{i+1} = f(z_i) \quad \text{if} \quad \Re e\, z_i < 0.$$

Then we have $z_{P+Q} = f^Q(z_0) - P = z_0$. On the other hand, let $z'_i = H(z_i)$; we have

$$z'_{i+1} = z'_i - 1 \quad \text{if} \quad z_{i+1} = f(z_i),$$
$$z'_{i+1} = F(z'_i) \quad \text{if} \quad z_{i+1} = z_i - 1,$$

hence $z'_{P+Q} = f^P(z'_0) - Q = z'_0$. Because H is orientation reversing on the equators, the periodic orbit $\mathcal{O}'$ in 4.6.2 (iii) will actually be the orbit of $\bar{z}'_0$.

To estimate $\Im m\, z'_i$, $0 \leq i < P + Q$, we consider H in the ball B of center $x_i = \Re e\, z_i$ and radius $\frac{1}{2}(\Delta - c_{23})$; the functional equation $H(f(z)) = H(z) - 1$ allows to define H in this ball and to check that it is univalent there.

In 4.6.2, we take c_{15} large enough and c_{16} small enough.

If $\Delta \leq c_{16}^{-1}\alpha$, the image by H of the interval $B \cap \mathbb{R}$ of length $\Delta - c_{23}$ by the functional equation will be of length at most $2c_{16}^{-1}$; by Koebe's 1/4-theorem, we will have $DH(x_i) \leq c'_{20}\Delta^{-1}$, and by the standard distorsion estimates we will obtain $|\Im m\, z'_i| < c_{20}\Delta^{-1}\Im m\, z_i$.

If $\Delta > c_{16}^{-1}\alpha$, we use $H(f(x_i)) = H(x_i) - 1$ and the standard distorsion estimates to obtain $|DH(x_i)| \leq \frac{5}{4}\alpha^{-1}$ (with c_{16} small enough), and then (with c_{15} large enough) $|\Im m\, z'_i| < \frac{4}{3}\alpha^{-1}\Im m\, z_i$.

This concludes the proof of Proposition 4.6.2 and of the construction of non linearizable analytic diffeomorphisms.

5 Appendix: Estimates of moduli of annular domains

5.1 Dirichlet integrals

Let A be an annular Riemann surface with boundary $\partial A = \partial_0 A \sqcup \partial_1 A$. Let $E(A)$ be the space of functions $H \in W^{1,2}(A)$ such that $H|_{\partial_0 A} \equiv 0$, $H|_{\partial_1 A} \equiv 1$. For $H \in E(A)$ define the Dirichlet integral

$$D(H) = \int_A |\nabla H|^2.$$

Then, we have

$$(\mathrm{mod} A)^{-1} = \inf_{E(A)} D(H).$$

In particular, for any $H_0 \in E(A)$

$$\mathrm{mod} A \geq D(H_0)^{-1}.$$

The annular Riemann surfaces for which we need to bound from below the modulus were obtained in two similar but slightly distincts ways.

5.2 First kind of moduli estimates

In the first setting, we have an holomorphic map $f(z) = z + \varphi(z)$. A domain U has for boundary a vertical segment $\ell = [ih_0, ih_1]$, its image $f(\ell)$ and the segments $[ih_0, f(ih_0)]$, $[ih_1, f(ih_1)]$. We obtain an annular Riemann surface A when we glue ℓ and $f(\ell)$ through f.

For $t \in [0, 1]$, $h \in [h_0, h_1]$, set

$$x + iy = ih + t\varphi(ih),$$
$$\varphi_1(h) = \mathfrak{Re}\,\varphi(ih),$$
$$\varphi_2(h) = \mathfrak{Im}\,\varphi(ih).$$

One has

$$dx = \varphi_1(h)dt + tD\varphi_1(h)dh$$
$$dy = \varphi_2(h)dt + (1 + tD\varphi_2(h))dh,$$

hence

$$dh = [\varphi_1(h)dy - \varphi_2(h)dx][\varphi_1 + t(\varphi_1 D\varphi_2 - \varphi_2 D\varphi_1)]^{-1},$$
$$dx \wedge dy = [\varphi_1 + t(\varphi_1 D\varphi_2 - \varphi_2 D\varphi_1)]dt \wedge dh.$$

Set now

$$H(x, y) = \frac{h - h_0}{h_1 - h_0}.$$

We will have $H \in E(A)$ and

$$D(H) = (h_1 - h_0)^{-2} \int |\nabla h|^2 \tag{5.1}$$

$$= (h_1 - h_0)^{-2} \int \int (\varphi_1^2 + \varphi_2^2)[\varphi_1 + t(\varphi_1 D\varphi_2 - \varphi_2 D\varphi_1)]^{-1} dt dh \tag{5.2}$$

$$\leq (h_1 - h_0)^{-2} \int_{h_0}^{h_1} \frac{|\varphi(ih)|^2 dh}{\varphi_1(h) - |\varphi_1(h) D\varphi_2(h)| - |\varphi_2(h) D\varphi_1(h)|}. \tag{5.3}$$

5.2.1 Estimates for $\tilde{A}_y$, $\tilde{A}_y^+$ in 4.2.2 We have there, with

$$h_{\max} = \Delta - \frac{1}{2\pi} \log \alpha_0^{-1} - c_8$$

the following estimates for $0 \leq h \leq h_{\max}$:

$$|\varphi(ih) - \alpha_0| \leq c_5 \alpha_0 \exp(-2\pi c_8) \exp[-2\pi(h_{\max} - h)],$$
$$|D\varphi(ih)| \leq c_6 \alpha_0 \exp(-2\pi c_8) \exp[-2\pi(h_{\max} - h)].$$

Plugging in these estimates in (5.1), we obtain

$$D(H) \leq (h_1 - h_0)^{-1}\alpha_0 + c_9'(h_1 - h_0)^{-2}\alpha_0$$

and then

$$(D(H))^{-1} \geq \alpha_0^{-1}(h_1 - h_0)[1 - c_9''(h_1 - h_0)^{-1}]$$

which gives the required estimates for the moduli of $\tilde{A}_y$, $\tilde{A}_y^+$.

5.2.2 Estimate for $\bar{V}$ in 4.5.1 One takes $h_0 = 0$, $h_1 = \frac{1}{3}\theta \exp(2\pi\Delta)|f(0)|$; one has

$$|f(z) - z - f(0)| \leq \frac{1}{10}|f(0)|, \tag{4.27}$$

for $|\Re z| \leq 3|f(0)|$, $|\Im m \, z| \leq h$, and

$$|Df(z) - 1| \leq c_6' \exp(-2\pi\Delta) \tag{4.23}$$

for $|\Im m \, z| \leq 1$. Plugging in these estimates in (5.1) gives the required estimate:

$$\text{mod}\bar{V} \geq c_{13}\theta \exp(2\pi\Delta).$$

5.2.3   Estimates for the punctures of $\bar{U}$ in 4.6.4 The glueing there is made through $g(z) = z + \alpha' + \psi(z)$, for $z = ih$, $h \geq \Delta - c_{23} + 1$, with

$$|\psi(z)| \leq \frac{1}{10} \exp[-2\pi(\Im m\, z - \Delta + c_{23} - 1)],$$

$$|D\psi(z)| \leq \frac{1}{10} \exp[-2\pi(\Im m\, z - \Delta + c_{23} - 1)].$$

From (5.1), we immediately get that the corresponding annulus has infinite modulus.

5.2.4   Estimate for the modulus of A_2' in 4.6.4 The glueing is made through f, which satisfies

$$|f(z) - z - \alpha| \leq \frac{1}{10} \exp[-2\pi(\Delta - c_{23} - \Im m\, z)],$$

$$|Df(z) - 1| \leq \frac{1}{10} \exp[-2\pi(\Delta - c_{23} - \Im m\, z)],$$

with $h_0 = 0$ and $h_1 = \Delta - c_{23} - \frac{\alpha}{2}$. It is assumed that $\Delta > c_{15}$, $\Delta - c_{23} \geq \alpha$. We obtain in (5.1):

$$D(H) \leq (h_1 - h_0)^{-2}[\alpha(h_1 - h_0) + \tilde{c}_{17}]$$

which gives the required estimate for A_2':

$$\mathrm{mod}\, A_2' \geq \Delta \alpha^{-1} - c_{17}''.$$

5.3   Second kind of moduli estimates

The second setting is the following. On one side, we have an exactly semi-circular domain

$$A_L = \{r_0 < |z - \omega| < r_1, \Re e(z - \omega) > 0\}$$

and on the other an approximately semicircular domain A_R, whose boundary is made from

- the two arcs $f^{\pm}([\omega \pm ir_0, \omega \pm ir_1])$;
- the two left half-circles with diameters $[f^{-}(\omega - ir_j), f^{+}(\omega + ir_j)]$, $j = 0, 1$;

the maps f^{-}, f^{+} here are close to a translation by some real number $\alpha > 1$.
    We glue the "vertical" sides of A_L, A_R through f^{+}, f^{-} to obtain A. For $r_0 \leq r \leq r_1$, $0 \leq t \leq \frac{1}{2}$, we set

$$z(t, r) = x(t, r) + iy(t, r) = \omega + \frac{1}{i} \exp(2\pi it)r.$$

On the other hand, for $1/2 \leq t \leq 1$, $r_0 \leq r \leq r_1$, we set

$$z(t,r) = x(t,r) + iy(t,r) = f_c(r) + \frac{1}{i}\rho(r)\exp(2\pi it),$$

with

$$f_c(r) = \frac{1}{2}[f^-(\omega - ir) + f^+(\omega + ir)]$$

$$\rho(r) = \frac{1}{2i}[f^+(\omega + ir) - f^-(\omega - ir)].$$

The formula

$$H(z(t,r)) = \frac{\log r/r_0}{\log r_1/r_0}$$

defines a function in $E(A)$, with

$$D(H) = \left(\log\frac{r_1}{r_0}\right)^{-2} \int_A |\nabla(\log r)|^2$$

$$\int_{A_L} |\nabla(\log r)|^2 = \int_{r_0}^{r_1} \pi\frac{dr}{r} = \pi\log\frac{r_1}{r_0}.$$

In A_R, we write

$$dx + idy = a\,dr + b\,dt$$

with

$$a = a_0 + ia_1 = Df_c(r) + \frac{1}{i}D\rho(r)\exp(2\pi it),$$

$$b = b_0 + ib_1 = 2\pi\rho(r)\exp(2\pi it),$$

giving

$$\int_{A_R} |\nabla(\log r)|^2 = \int_{1/2}^1 \int_{r_0}^r \frac{|b|^2}{|a_0b_1 - a_1b_0|}dt\,dr.$$

Now, we have

$$\frac{|b|^2}{|a_0b_1 - a_1b_0|} \leq 2\pi\left[\Re e\left(\frac{D\rho}{\rho}\right) - \frac{|Df_c|}{\rho}\right]^{-1}$$

giving

$$\int_{A_R} |\nabla(\log r)|^2 \leq \pi \int_{r_0}^{r_1}\left[\Re e\left(\frac{D\rho}{\rho}\right) - \frac{|Df_c|}{\rho}\right]^{-1} dr. \qquad (5.4)$$

5.3.1   Estimates for A_1, A_2 in 4.6.4 We take $f^+ = g$, $f^- = f$, $r_0 = 3$, $\omega = i\left(\Delta - c_{23} + \frac{1}{2}\right)$ and

$$r_1 = \begin{cases} \Delta - c_{23} - 1 & \text{if } 5 \leq \Delta - c_{23} < \frac{\alpha}{2} \\ \frac{\alpha}{2} - 1 & \text{if } \Delta - c_{23} \geq \frac{\alpha}{2} \geq 5. \end{cases}$$

We have then, from the estimates on f, g in 4.6.4

$$\left| \rho(r) - \left(\frac{\alpha - \alpha'}{2i} + r \right) \right| \leq \frac{1}{10} \exp\left[-2\pi \left(r - \frac{1}{2} \right) \right],$$

$$|D\rho(r) - 1| \leq \frac{1}{10} \exp\left[-2\pi \left(r - \frac{1}{2} \right) \right],$$

$$|Df_c(r)| \leq \frac{1}{10} \exp\left[-2\pi \left(r - \frac{1}{2} \right) \right].$$

When we plug these estimates in (5.4), we obtain

$$\int_{A_R} |\nabla(\log r)|^2 \leq \pi \log \frac{r_1}{r_0} + c''_{18}$$

giving the required estimates for $\mathrm{mod}\,A_1$, $\mathrm{mod}\,A_2$ in 4.6.4.

5.3.2   Estimate for A_3 in 4.6.4 We take now (with $c_{16}\Delta \leq \alpha$, c_{16} small) $\omega = 0$, $f^+ = g$, $f^- = \bar{g}$, $r_0 = \Delta - c_{23} + 2$, $r_1 = \frac{\alpha}{2} - 1$. The same computation as above will give

$$\int_{A_R} |\nabla(\log r)|^2 \leq \pi \log \frac{r_1}{r_0} + c''_{18}$$

and the required estimate for $\mathrm{mod}\,A_3$.

Acknowledgements. I am extremely grateful to Timoteo Carletti, who wrote a first version of these notes, which served as a basis to the present manuscript. I would also like to thank very much Stefano Marmi, whose friendly pressure and constant help finally allowed this text to exist.

References

[Ah]   L.V. Ahlfors: Lectures on Quasiconformal Mappings, Van Nostrand (1966)

[Ar1]   V.I. Arnol'd: Small denominators I. On the mapping of a circle into itself. Izv. Akad. Nauk. Math. Series **25**, 21–86 (1961) [Transl. A.M.S. Serie 2, **46**, (1965)]

[DMVS] W. De Melo and S.J. Van Strien: One-dimensional dynamics, Springer-Verlag, New York, Heidelberg, Berlin (1993)

[De]   A. Denjoy: Sur les courbes définies par les équations differentielles à la surface du tore, J. Math. Pures et Appl. **11**, série 9, 333–375, (1932)

[HW]   G.H. Hardy and E.M. Wright: An introduction to the theory of numbers, 5th edition Oxford Univ. Press

[He1]   M.R. Herman: Sur la conjugaison differentielle des difféomorphismes du cercle à des rotations. Publ.Math. I.H.E.S. **49**, 5–234 (1979)

[He2]   M.R. Herman: Simple proofs of local conjugacy theorems for diffeomorphisms of the circle with almost every rotation number, Bol. Soc. Bras. Mat. **16**, 45–83 (1985)

[KO1]   Y. Katznelson and D. Ornstein: The differentiability of the conjugation of certain diffeomorphisms of the circle, Ergod. Th. and Dyn. Sys. **9**, 643–680 (1989)

[KO2]   Y.Katznelson and D.Ornstein: The absolute continuity of the conjugation of certain diffeomorphisms of the circle, Ergod. Th. and Dyn. Sys. **9**, 681–690 (1989)

[KS]   K.M. Khanin and Ya. Sinai: A new proof of M. Herman's theorem, Commun. Math. Phys. **112**, 89–101 (1987)

[MMY1]   S. Marmi, P. Moussa and J.-C. Yoccoz: The Brjuno functions and their regularity properties, Communications in Mathematical Physics **186**, 265–293 (1997)

[MMY2]   S. Marmi, P. Moussa and J.-C. Yoccoz: Complex Brjuno functions, Journal of the American Mathematical Society, **14**, 783–841 (2001)

[PM]   R. Pérez-Marco: Fixed points and circle maps, Acta Math. **179**, 243–294 (1997)

[Ri]   E. Risler: Linéarisation des perturbations holomorphes des rotations et applications, Mémoires S.M.F. **77** (1999)

[SK]   Ya.G. Sinai and K.M. Khanin: Smoothness of conjugacies of diffeomorphisms of the circle with rotations, Russ. Math. Surv. **44**, 69–99 (1989)

[Yo1]   J.-C. Yoccoz: Conjugaison differentielle des difféomorphismes du cercle dont le nombre de rotation vérifie une condition Diophantienne, Ann. Sci. Ec. Norm. Sup. **17**, 333–361(1984)

[Yo2]   J.-C. Yoccoz: Théorème de Siegel, polynômes quadratiques et nombres de Brjuno, Astérisque **231**, 3–88 (1995)

Some open problems related to small divisors

S. Marmi, J.-C. Yoccoz

0   Introduction

What follows is a redaction of a (memorable) three–hours–long open problem session which took place during the workshop in Cetraro. Each of the five lecturers (H. Eliasson, M. Herman, S. Kuksin, J.N. Mather and J.-C. Yoccoz) spent about half an hour briefly introducing some open problems. This redaction grew from the notes that the first author took during the session: whereas we mention who suggested each of the problems of the list we give below (with the exception of the authors of this text) we are the only responsible for any mistake the reader may find in their description or formulation. Moreover the list of references is very far from being complete and has not been updated. We tried to make this text self-contained, but see [KH] for terminology and further information and [Yo2] for a short survey of classical results concerning small divisor problems.

1   One-Dimensional Small Divisor Problems (On Holomorphic Germs and Circle Diffeomorphisms)

1.1   Linearization of the quadratic polynomial. Size of Siegel disks

Let us consider the linearization problem for the quadratic polynomial $P_\lambda(z) = \lambda(z - z^2)$ ([Yo3], Chapter II) where $z \in \mathbb{C}$, $\lambda = e^{2\pi i \alpha}$ and $\alpha \in \mathbb{C}/\mathbb{Z}$. We say that P_λ is *linearizable* if there exists a holomorphic map tangent to the identity $h_\lambda(z) = z + \mathcal{O}(z^2)$ such that $h_\lambda(\lambda z) = P_\lambda(h_\lambda(z))$. Then h_λ is unique and we will denote r_λ its radius of convergence (when $|\lambda| = 1$ this measures the "size" of the Siegel disk of P_λ).

The second author proved the following results:

(1) there exists a bounded holomorphic function $U : \mathbb{D} \to \mathbb{C}$ such that for all $\lambda \in \mathbb{D}$, $|U(\lambda)|$ is equal to r_λ;
(2) for all $\lambda_0 \in \mathbb{S}^1$, $|U(\lambda)|$ has a non-tangential limit in λ_0, which is still equal to r_{λ_0};
(3) if $\lambda = e^{2\pi i \alpha}$, $\alpha \in \mathbb{R} \setminus \mathbb{Q}$, P_λ is linearizable if and only if α is a Brjuno number: if $(p_n/q_n)_{n \geq 0}$ denotes the sequence of the convergents of the continued fraction expansion of α then being a *Brjuno number* means that $\sum_{n=0}^\infty \frac{\log q_{n+1}}{q_n} < +\infty$;

(4) There exists a universal constant $C_1 > 0$ and for all $\varepsilon > 0$ there exists
$C_\varepsilon > 0$ such that for all Brjuno numbers α one has

$$(1 - \varepsilon)B(\alpha) - C_\varepsilon \leq -\log|U(e^{2\pi i\alpha})| \leq B(\alpha) + C_1 .$$

Problem 1.1.1 *Does the function (defined on the set of Brjuno numbers)*
$\alpha \mapsto B(\alpha) + \log|U(e^{2\pi i\alpha})|$ *belong to* $L^\infty(\mathbb{R}/\mathbb{Z})$?

There is a good numerical evidence [Mar] in support to a positive answer
to the following much stronger property:

Problem 1.1.2 *Does the function* $\alpha \mapsto B(\alpha) + \log|U(e^{2\pi i\alpha})|$ *extend to a*
1/2-Hölder continuous function as α *varies in* $\mathbb{R}$?

These two problems are discussed to some extent in [MMY1], [MMY2].
For some related analytical and numerical results concerning some area-
preserving maps, including the standard family, we refer to [Mar], [BPV],
[MS], [Da], [BG], [CL].

1.2   Herman rings. Differentiable conjugacy of diffeomorphisms of the circle

The second author [Yo5] proved that the Brjuno condition is also necessary
and sufficient in the local conjugacy problem for analytic diffeomorphisms
of the circle. In this case the simplest non-trivial model is provided by the
Blaschke products $Q_{a,\alpha}(z) = \rho_\alpha z^2 \frac{z+a}{1+az}$. Here $a \in (3, +\infty)$ and $\rho_\alpha \in \mathbb{S}^1$ is
chosen in such a way that the rotation number of the restriction of $Q_{a,\alpha}$
to $\mathbb{S}^1$ is exactly α. Under these assumptions $Q_{a,\alpha}$ induces an orientation-
preserving analytic diffeomorphism of $\mathbb{S}^1$. Note that when $a \to +\infty$ then
$Q_{a,\alpha}(z) \to e^{2\pi i\alpha}z$.

When α is a Brjuno number if a is large enough then $Q_{a,\alpha}$ is analytically
conjugated to the rotation $R_\alpha(z) = e^{2\pi i\alpha}z$ in a neighborhood of $\mathbb{S}^1$. If α
satisfies the more restrictive arithmetical condition $\mathcal{H}$ (we refer to Yoccoz's
lectures in this volume for its definition) then $Q_{a,\alpha}$ is conjugated to the
rotation for all $a > 3$. This leads to the following

Problem 1.2.1 *Let* α *be a Brjuno number not satisfying condition* $\mathcal{H}$: *does*
there exist an $a > 3$ *such that* $Q_{a,\alpha}$ *is not analytically conjugated to the*
rotation R_α ?

Concerning this problem Herman showed that there exists at least a Brjuno
number α not satisfying $\mathcal{H}$ such that the answer to the previous question is
positive.

In general one expects to exist a maximal interval $(a_0, +\infty)$, $a_0 > 3$, such
that $Q_{a,\alpha}$ is analytically conjugated to a rotation for all $a \in (a_0, +\infty)$ whereas
$Q_{a_0,\alpha}$ is not analytically conjugated.

Problem 1.2.2 *How smooth is the conjugacy for* a_0 ? *How does this smooth-*
ness depend on α ?

This leads naturally to the study of conjugacy classes of orientation-preserving diffeomorphisms of the circle with finitely many continuous derivatives. Here the relevant arithmetical conditions on the rotation number are of Diophantine type. Let $\tau \geq 0$; we denote $DC(\tau)$ the set of irrational numbers α whose denominators q_n of the continued fraction expansion satisfy $q_{n+1} = \mathcal{O}(q_n^{1+\tau})$ for all $n \geq 0$.

Let $\mathbb{T} = \mathbb{R}/\mathbb{Z}$; $r, s \in \{0, +\infty, \omega\} \cup [1, +\infty)$. Let $\mathrm{Diff}_+^r(\mathbb{T})$ be the group of C^r diffeomorphisms[1] of $\mathbb{T}$ which are orientation-preserving. We denote $D^r(\mathbb{T})$ the group of C^r diffeomorphisms of the real line such that $\tilde{f} - \mathrm{id}$ is $\mathbb{Z}$-periodic. We consider the linearization problem $f \circ h = h \circ R_\alpha$ where R_α denotes the rotation of α on $\mathbb{T}$, α is the rotation number of f (mod 1) and $h \in \mathrm{Diff}_+^s(\mathbb{T})$, with $r \geq s$. One must distinguish the *local* conjugacy problem from the *global* one: thus we define

$$C_{r,s} = \{\alpha \in \mathbb{R} \setminus \mathbb{Q}, \text{ every } f \in \mathrm{Diff}_+^r(\mathbb{T}) \text{ with rotation number } \alpha \bmod 1$$
$$\text{is conjugated to } R_\alpha \text{with a conjugacy } h \in \mathrm{Diff}_+^s(\mathbb{T})\}$$

$$C_{r,s}^{\mathrm{loc}} = \{\alpha \in \mathbb{R} \setminus \mathbb{Q}, \text{ every } f \in \mathrm{Diff}_+^r(\mathbb{T}) \text{ with rotation number } \alpha \bmod 1$$
$$C^r\text{-close to } R_\alpha \text{ is conjugated to } R_\alpha \text{ with a conjugacy } h \in \mathrm{Diff}_+^s(\mathbb{T})\}$$

Note that in the definition of $C_{r,s}^{\mathrm{loc}}$ the neighborhood in the C^r topology which measures the distance of f from R_α depends on α.

Let $s < r - 1 < \infty$. The following inclusions are known after [He2, KO1, KO2, KS, Yo1, Yo4], etc.

$$DC(r - s - 1 - \varepsilon) \subset C_{r,s} \subset C_{r,s}^{\mathrm{loc}} \subset DC(r - s - 1 + \varepsilon) \text{ for all } \varepsilon > 0$$

The third inclusion is due to Herman [He2]. One also knows from [SK] that:

- if $1 < s < 2 < s + 1 < r < 3$ then $DC(r - s - 1) \subset C_{r,s}$;
- $DC(0) \subset C_{r,r-1}$ provided that $r > 2$, $r \notin \mathbb{N}$.

Problem 1.2.3 *Determine $C_{r,s}$ and $C_{r,s}^{\mathrm{loc}}$. Are they different ?*

1.3 Gevrey classes

In the case of the conjugacy of germs of formal diffeomorphisms of $(\mathbb{C}, 0)$ one can consider a problem similar to the local one above requiring the formal germs to belong to some ultradifferentiable class, for example Gevrey classes.

Consider two subalgebras $A_1 \subset A_2$ of $z\mathbb{C}[[z]]$ closed with respect to the composition of formal series. In addition to the usual cases $z\mathbb{C}[[z]]$ (formal germs) and $z\mathbb{C}\{z\}$ (analytic germs) one can for example consider Gevrey-s classes G_s, $s > 0$ (i.e. series $F(z) = \sum_{n \geq 0} f_n z^n$ such that there exist

---

[1] If $r = 0$ it is the group of homeomorphisms of $\mathbb{T}$; if $r \geq 1$, $r \in \mathbb{R} \setminus \mathbb{N}$, it is the group of $C^{[r]}$ diffeomorphisms whose $[r]$-th derivative satisfies a Hölder condition of exponent $r - [r]$; if $r = \omega$ it is the group of $\mathbb{R}$-analytic diffeomorphisms.

$c_1, c_2 > 0$ such that $|f_n| \leq c_1 c_2^n (n!)^s$ for all $n \geq 0$). Let $F \in A_1$ being such that $F'(0) = \lambda \in \mathbb{C}^*$. We say that F is linearizable in A_2 if there exists $H \in A_2$ tangent to the identity and such that $F \circ H = H \circ R_\lambda$ where $R_\lambda(z) = \lambda z$. Let $\lambda = e^{2\pi i \alpha}$ with $\alpha \in \mathbb{R} \setminus \mathbb{Q}$.

One knows that if α is a Brjuno number then for all $s > 0$ all germs $F \in G_s$ have a linearization $H \in G_s$ (see [CM]). Let $r > s > 0$ and denote

$$\mathcal{G}_{r,s} = \{\alpha \in \mathbb{R}\setminus\mathbb{Q} \ , \ \text{every } F \in G_s \text{ is conjugated to } R_\alpha \text{with a conjugacy} H \in G_r\}$$

One knows that a condition weaker than Brjuno is sufficient [CM].

Problem 1.3.1 *Determine* $\mathcal{G}_{r,s}$.

Of course one can ask a similar question in the circle case, distinguishing the local from the global case.

2   Finite-Dimensional Small Divisor Problems

2.1   Linearization of germs of holomorphic diffeomorphisms of $(\mathbb{C}^n, 0)$

Let $n \geq 2$ and let $f \in (\mathbb{C}[[z_1, \ldots z_n]])^n$ be a germ of formal diffeomorphism of $(\mathbb{C}^n, 0)$, $z = (z_1, \ldots, z_n)$, $f(z) = Az + O(z^2)$ with $A \in \mathrm{GL}(n, \mathbb{C})$. Let $\lambda_1, \ldots, \lambda_n$ denote the eigenvalues of A, $k = (k_1, \ldots, k_n) \in \mathbb{N}^n$, $\lambda^k = \lambda_1^{k_1} \ldots \lambda_n^{k_n}$ and $|k| = \sum_{j=1}^n |k_j|$. Assume that the eigenvalues are all distinct. If

$$\lambda^k - \lambda_j \neq 0 \text{ for all } j = 1, \ldots, n \text{ and } k \in \mathbb{N}^n, \ |k| \geq 2 \qquad \text{(NR)}$$

then f is formally linearizable, i.e. there exists a unique germ h of formal diffeomorphism of $(\mathbb{C}^n, 0)$, tangent to the identity, such that $h^{-1} \circ f \circ h = A$.

Let f be $\mathbb{C}$-analytic and assume that A satisfies (NR). For $m \in \mathbb{N}$, $m \geq 2$ let

$$\Omega(m) = \inf_{2 \leq |k| \leq m} \inf_{1 \leq j \leq n} |\lambda^k - \lambda_j| \ .$$

Then Brjuno [Br] proved that if A is diagonalizable, satisfies (NR) and the condition

$$\sum_{k=0}^\infty 2^{-k} \log(\Omega(2^{k+1}))^{-1} < +\infty \qquad \text{(B)}$$

then f is analytically linearizable, i.e. the formal germ h defines a germ of $\mathbb{C}$-analytic diffeomorphism of $(\mathbb{C}^n, 0)$. The proof uses the classical majorant series method used by Siegel [S, St] to prove that h is $\mathbb{C}$-analytic under the stronger assumption that $\lambda_1, \ldots, \lambda_n$ satisfy a diophantine condition.

Problem 2.1.1 (M. Herman) *What is the optimal arithmetical condition on the eigenvalues of the linear part which assures that f is analytically linearizable? Can one obtain it by direct majorant series method?*

It seems unlikely that condition (B) is optimal.

Concerning the problem of linearization of germs of holomorphic diffeomorphisms near a fixed point [He4] contains many other questions, most of which are still open.

2.2   Elliptic fixed points and KAM theory

If one replaces the assumption of being conformal with the assumption of preserving the standard symplectic structure of $\mathbb{R}^{2n}$ one can consider the following problem.

Let f be a real analytic symplectic diffeomorphism of $\mathbb{R}^{2n}$ which leaves the origin fixed $f(0) = 0$. Let $z \in \mathbb{R}^{2n}$ and assume that $f(z) = Az + O(z^2)$, where $A \in \mathrm{Sp}\,(2n, \mathbb{R})$ is conjugated in $\mathrm{Sp}\,(2n, \mathbb{R})$ to $r_{\alpha_1} \times \ldots \times r_{\alpha_n}$, where

$$r_{\alpha_i} = \begin{pmatrix} \cos 2\pi\alpha_i & \sin 2\pi\alpha_i \\ -\sin 2\pi\alpha_i & \cos 2\pi\alpha_i \end{pmatrix} \text{ and the vector } \alpha = (\alpha_1, \ldots, \alpha_n) \in \mathbb{R}^n \text{ satisfies}$$

a diophantine condition. Herman [He6] stated the following conjecture.

Problem 2.2.1 (M. Herman) *Show that there exists $\varepsilon_0 > 0$ such that in the ball $\|z\| \leq \varepsilon_0$ there is a set of positive Lebesgue measure of invariant Lagrangian tori.*

Note that here the assumption of f being real analytic is essential since in the C^∞ case the conjecture is true for $n = 1$, open if $n = 2$ but one knows the existence of counterexamples if $n \geq 3$. Using Birkhoff normal form one can prove that the conjecture is true in many special cases (some twist condition, see [BHS]).

2.3 $\mathbb{Z}^k$-actions

Let $k \geq 1$ and $F_1, \ldots, F_k$ be commuting diffeomorphisms in $D^r(\mathbb{T})$. Let $\alpha_1, \ldots, \alpha_k$ be the rotation numbers of $F_1, \ldots, F_k$. Then the diffeomorphisms $f_1, \ldots, f_k$ of the circle induced by $F_1, \ldots, F_k$ generate a $\mathbb{Z}^k$-action on $\mathbb{T}$. If α_1 is irrational and $F_1 = R_{\alpha_1}$, then we must have $F_i = R_{\alpha_i}$ for $1 \leq i \leq k$, because the centralizer of R_{α_1} in $D^0(\mathbb{T})$ is the group of translations. Therefore, if α_1 is irrational and F_1 is conjugated to R_{α_1} by a diffeomorphism $h \in D^s(\mathbb{T})$, the full $\mathbb{Z}^k$-action is linearized by h.

J. Moser [M2] has shown that, if for some γ, τ and all $q > 0$

$$\max_{1 \leq i \leq k} \|q\alpha_i\| \geq \gamma q^{-\tau}$$

then there exists a neighborhood V of the identity in $D^\infty(\mathbb{T})$ such that if $F_i \in D^\infty(\mathbb{T})$, $\rho(F_i) = \alpha_i$ and $F_i \circ R_{-\alpha_i} \in V$ then the action is linearizable in $D^\infty(\mathbb{T})$. The simultaneous approximation condition above is probably optimal in the C^∞ category.

More generally, one can define, for $1 \leq s \leq r \leq \infty$, or $r = s = \omega$

$$C_{r,s}^k = \{(\alpha_1, \ldots, \alpha_k), \text{ any } k\text{-uple}\,(F_1, \ldots, F_k)$$

of commuting diffeomorphisms in

$(D^r(\mathbb{T}))^k$ with rotation numbers $(\alpha_1, \ldots, \alpha_k)$ is linearizable in $D^s(\mathbb{T})\}$,

and similarly $C_{r,s}^{k,loc}$ if one assumes furthermore that $F_i \circ R_{-\alpha_i}$, $1 \leq i \leq k$, are C^r-close to the identity.

Problem 2.3.1 *Determine* $C_{r,s}^k$, $C_{r,s}^{k,loc}$.

Progress in this direction is contained in the papers [DL], [Kra], [PM], etc.

The previous problem generalizes to the study of $\mathbb{Z}^k$ actions on $\mathbb{R}^n$. Let $k > n \geq 1$ and consider an action $\mathbb{Z}^k \hookrightarrow \text{Diff}_+^\omega(\mathbb{R}^n)$. Among them the actions by translations $R_{\alpha_i} : x \mapsto x + \alpha_i$, $1 \leq i \leq k$, where $\alpha_i \in \mathbb{R}^n$, play a distinguished role. Assume that $\alpha_1, \ldots, \alpha_k$ generate $\mathbb{R}^n$ and consider those actions whose generators $f_1, \ldots, f_k \in \text{Diff}_+^\omega(\mathbb{R}^n)$ are C^ω-close to $R_{\alpha_1}, \ldots, R_{\alpha_k}$ and which are C^0-conjugated to it (thus one can call $\alpha_1, \ldots, \alpha_k$ the rotation numbers of the action.

Problem 2.3.2 *For which rotation numbers is this C^0-conjugacy indeed analytic?*

2.4 Diffeomorphisms of compact manifolds

Let M be a C^∞ compact connected manifold. We denote $\text{Diff}^\infty(M)$ the group of C^∞ diffeomorphisms of M with the C^∞ topology and $\text{Diff}_+^\infty(M)$ the group of C^∞ diffeomorphisms C^∞-isotopic to the identity (i.e. the connected component of $\text{Diff}^\infty(M)$ which contains the identity).

The general problem that one may address is to study the structure (conjugacy classes, centralizers, etc.) of $\text{Diff}^\infty(M)$. In the special case of the n-dimensional torus $\mathbb{T}^n = \mathbb{R}^n/\mathbb{Z}^n$ one has a KAM theorem which describes the local structure of $\text{Diff}_+^\infty(\mathbb{T}^n)$ near diophantine translations $R_\alpha : x \mapsto x + \alpha$, $\alpha \in \mathbb{T}^n$: there exists a neighborhood U_α of R_α in $\text{Diff}_+^\infty(\mathbb{T}^n)$ such that if $f \in U_\alpha$ there exists $\lambda \in \mathbb{T}^n$ and $g \in \text{Diff}_+^\infty(\mathbb{T}^n)$ such that $g(0) = 0$ and $f = R_\lambda \circ g^{-1} \circ R_\alpha \circ g$. Moreover this decomposition is locally unique.

In [He1] a "converse" of this theorem is asked:

Problem 2.4.1 (M. Herman) *Let V be a compact C^∞ manifold, $f \in \text{Diff}^\infty(V)$, U a C^∞ neighborhood of the identity, $\mathcal{O}_{f,U} = \{g \circ f \circ g^{-1}, \ g \in U\}$. If $\mathcal{O}_{f,U}$ is a finite codimension manifold is it true that $V = \mathbb{T}^n$ and f is C^∞-conjugate to a Diophantine translation ?*

In the torus case one proves KAM theorem by means of an implicit function theorem in Fréchet spaces (see [Ha, Bo]). The main point is that a translation R_α being diophantine is equivalent to ask that for all $\varphi \in C^\infty(\mathbb{T}^n)$ there exist $\psi \in C^\infty(\mathbb{T}^n)$ and $\lambda \in \mathbb{R}$ such that the linearized conjugacy equation

$$\psi \circ R_\alpha - \psi = \varphi + \lambda$$

holds. Then one can ask the analogue of Problem 2.4.1 for the linearized conjugacy equation

Problem 2.4.2 (M. Herman) *If for all $\varphi \in C^\infty(V)$ there exists $\psi \in C^\infty(V)$ and $\lambda \in \mathbb{R}$ such that $\psi \circ f - \psi = \varphi + \lambda$ is it true that $V = \mathbb{T}^n$ and f is C^∞-conjugate to a Diophantine translation ?*

3   KAM Theory and Hamiltonian Systems

3.1   Twist maps

We let $\mathbb{T}$ denote the circle $\mathbb{R}/\mathbb{Z}$ and $\theta \pmod 1$ denote the standard parameter of $\mathbb{T}$ and x the corresponding parameter of its universal cover $\mathbb{R}$. We will let $y \in \mathbb{R}$ denote the standard parameter of the second factor of $\mathbb{T} \times \mathbb{R}$. Let U be an open subset of $\mathbb{T} \times \mathbb{R}$ which intersects each vertical line $\{\theta\} \times \mathbb{R}$ in an open non empty interval. We consider maps f which are diffeomorphisms from U onto an open subset $f(U) \subset \mathbb{T} \times \mathbb{R}$ which also intersects each vertical line in an open non empty interval. We assume that f is *orientation preserving* and *area preserving*. Since we are in two dimensions the area preserving condition is the same as requiring that f be *symplectic*. Let $\tilde{f}$ denote the lift of f to the universal cover so that $\tilde{f}(x+1,y) = \tilde{f}(x,y) + (1,0)$. We also set $(x',y') = \tilde{f}(x,y)$.

An orientation preserving symplectic C^1 diffeomorphism f satisfies a *positive* (resp. *negative*) *monotone twist* condition if $\frac{\partial x'}{\partial y} > \varepsilon$ (resp. $< -\varepsilon$) for some fixed $\varepsilon > 0$ and for all (x,y). Geometrically this condition states that the image of a segment $x =$ constant under $\tilde{f}$ forms a graph over the x axis. An *integrable* twist map has the form $\tilde{f}(x,y) = (x + r(y), y)$.

From the area preserving property of f it follows that $y'dx' - ydx$ is a closed 1-form and therefore there exists a *generating function* (or *variational principle*) $h = h(x,x')$ such that

$$y = -\partial_1 h(x,x') , \quad y' = \partial_2 h(x,x') .$$

The generating function is unique up to the addition of a constant and its invariance under translations $(x,x') \mapsto (x+1,x'+1)$ is equivalent to the condition that $y'dx' - ydx$ is exact on $\mathbb{T} \times \mathbb{R}$. Moreover, from the positive twist condition one has $\partial_{12}h(x,x') < 0$.

A *rotational* invariant curve is a homotopically non-trivial f-invariant curve. By Birkhoff's theory (see [He3], Chapitre I), such a curve is the graph of a Lipschitz function. For near-to-integrable twist maps KAM theory provides the existence of many rotational invariant curves. Herman [He3, He5] proved that rotational invariant curves persist in twist diffeomorphisms which are C^3-close to an integrable map.

Problem 3.1.1 (J. Mather) *Does there exist an example of a C^r twist area-preserving map with a rotational invariant curve which is not C^1 (separate question fo each $r \in [1,\infty] \cup \{\omega\}$).*

Problem 3.1.2 (J. Mather) *Given a C^∞ twist diffeomorphism and a rotational invariant curve which is not C^∞ is it possible to destroy it by an arbitrarily small C^∞ perturbation?*

One knows, following [Ma3] that if the rotation number is not diophantine this is indeed possible even in the case the circle is C^∞. [Ma3] contains also destruction results for C^r twist maps (see p. 212) and [Fo] for analytic maps but in both cases there is a gap between the destruction results and the persistence results given by KAM theory.

It is a classical counterexample of Arnold that there exist analytic diffeomorphisms of the circle, with irrational rotation number, whose conjugacy to a rigid rotation is not absolutely continuous. Since every diffeomorphism of the circle can be embedded as rotational invariant curve of an area–preserving monotone twist map of the annulus with the same degree of smoothness, this example, and those constructed by Denjoy [De1, De2] and Herman [He2] in the differentiable case, give examples of "regular" twist maps f having "regular" rotational invariant curves γ such that $f|_\gamma$ is topologically conjugate to a rigid rotation but the conjugacy is not absolutely continuous. In these examples the irrational rotation number has extremely good approximations by rational numbers.

The classical results of Denjoy on diffeomorphisms of the circle show that given an invariant curve γ, if the rotation number α of $f|_\gamma$ is irrational then one has the following:

- if $f|_\gamma \in C^2$ then $f|_\gamma$ is topologically conjugate to R_α and every orbit is dense in γ;
- there exist examples of $f|_\gamma \in C^{2-\varepsilon}$, $\varepsilon > 0$, such that no orbit is dense in γ and the limit set of the orbit of every point of γ is the same Cantor set (*Denjoy minimal set*).

Thus even if f is smooth, limit sets different from γ may appear provided that the invariant curve γ loses smoothness.

Problem 3.1.3 (J. Mather) *Does there exist a C^3 area–preserving twist map of the annulus with a rotational, not topologically transitive, invariant curve of irrational rotation number? Same problem with $r \geq 3$ or even analytic.*

M. Herman [He3] gave an example of class $C^{3-\varepsilon}$. Hall and Trupin [HT] gave a C^∞ example without the area–preserving condition.

The most important progress towards the understanding of these problems has come through the introduction of *Aubry–Mather sets* [AL, Ma1, Ma2, Ma4]. These are closed invariant sets given by a parametric representation

$$x = u(\theta), \quad y = -\partial_1 h(u(\theta), u(\theta + \alpha)),$$

where u is monotone (but not necessarily continuous) and $u - \theta$ is $\mathbb{Z}$-periodic. They do exist for all rotation numbers α and they are subsets of a closed Lipschitz graph. For rational α one obtains periodic orbits, whereas for irrational numbers one has rotational invariant curves if u is continuous (in fact Lipschitz by Birkhoff's theorem) or an invariant Cantor set if u has countably many discontinuities. In this case the Aubry–Mather set can be viewed as a Cantor set drawn on a Lipschitz graph. Another important property of Aubry–Mather sets is that the "ordering" of an orbit is the same as for the rotation by α of a circle.

Mather based his proof on the variational problem

$$\int_0^1 h(u(\theta), u(\theta + \alpha)) d\theta$$

minimizing this functional in the class of weakly monotone functions, thus they are also called *action minimising sets*.

Problem 3.1.4 (M. Herman) *For a C^r twist area-preserving diffeomorphism does the union of the action-minimising sets which are not closed curves and do not contain periodic orbits have Hausdorff dimension 0?*

Here $r \geq 3$; otherwise Herman himself has a counterexample.

3.2   Euler–Lagrange flows

It is proved in [M1] that any monotone twist map can be obtained as the time-1 map of the Hamiltonian flow associated to a time–dependent, $\mathbb{Z}$-periodic in time Hamiltonian $H : T^*(\mathbb{T} \times \mathbb{R}) \times \mathbb{R} \to \mathbb{R}$ satisfying Legendre condition $H_{yy}(\theta, y, t) > 0$. This assures that f can also be interpolated by the time-1 map of the Euler–Lagrange flow associated to the Lagrangian function $L : T(\mathbb{T} \times \mathbb{R}) \times \mathbb{R} \to \mathbb{R}$ obtained from H by Legendre transform.

More generally, let M be a a closed Riemannian manifold M and consider Lagrangians of the form kinetic energy + time periodic potential $V \in C^r(M \times \mathbb{T})$ (see [Ma5] for a more general setting). Assume that M has dimension at least 2.

Problem 3.2.1 (J. Mather) *Is there a residual set (in the sense of Baire category) in $C^r(M \times \mathbb{T})$ such that there exists a corresponding trajectory of the Euler–Lagrange flow with kinetic energy growing to ∞ as $t \to \infty$?*

Of course these systems are very far from integrable ones. De La Llave has some results concerning this problem.

3.3 n-body problem

Let $n \geq 2$. Consider $n + 1$ point masses $m_0, \ldots, m_n$ moving in an inertial reference frame $\mathbb{R}^3$ with the only forces acting on them being their mutual gravitational attraction. If the i-th particle has position vector q_i then the Newtonian equations of motion are

$$m_i \ddot{q}_i = -\frac{\partial V}{\partial q_i}, \quad i = 1, \ldots, n$$

where $V(q_0, \ldots, q_n) = -\sum_{0 \leq i < j \leq n} \frac{G m_i m_j}{|q_i - q_j|}$ and G is the universal gravitational constant. The size of the system is measured by the moment of inertia $I = \frac{1}{2} \sum_{i=0}^n m_i |q_i|^2$. The total energy of the system $E = T + V = \sum_{i=0}^n m_i \frac{|\dot{q}_i|^2}{2} + V(q_0, \ldots q_n)$, the total angular momentum $A = \sum_{i=0}^n m_i q_i \wedge \dot{q}_i$ and the total momentum $P = \sum_{i=0}^n m_i \dot{q}_i$ are integrals of the motion. We will assume that the center of mass $C = \sum_{i=0}^n m_i q_i$ is *fixed* at the origin, thus $P = 0$.

J. Mather recalled a problem proposed by G.D. Birkhoff [Bi]:

Problem 3.3.1 (G.D. Birkhoff) *Let $n = 2$ (three body problem) and assume that the total energy is negative. One can even assume that the three particles move on a fixed plane. Is it true that the moment of inertia of the system can grow to ∞ as $t \to \infty$ for a dense open set of initial conditions? One can ask the same question for $n > 2$.*

The restriction to negative energy is necessary since from the identity of Jacobi–Lagrange $\ddot{I} = E + T$ follows that if the energy $E \geq 0$ all orbits which are defined for all times are wandering. The only thing which is known is that even for negative energy wandering sets do exist (as Birkhoff and Chazy showed long ago, see [Al]).

The fact that the bounded orbits form a positive Lebesgue-measure set for the planar three-body problem is a consequence of KAM theory (see [Ar]). According to M. Herman [He6] "what seems not an unreasonable question to ask (and possibly prove in a finite time with a lot of technical details) is"

Problem 3.3.2 (M. Herman) *If one of the masses $m_0 = 1$ and all the other masses $m_j \ll 1$ are sufficiently small, are there wandering domains in any neighborhood of fixed distinct circular orbits around the mass m_0 and moving in the same direction in a plane?*

4   Linear Quasiperiodic Skew-Products, Spectral Theory and Hamiltonian Partial Differential Equations

4.1   Reducibility of skew-products

Let $\omega \in \mathbb{R}^d$, $\theta \in \mathbb{T}^d$ (here $\mathbb{T} = \mathbb{R}/2\pi\mathbb{Z}$), $V : \mathbb{T}^d \to \mathbb{R}$ be a continuous function. The *time-continuous one-dimensional quasiperiodic Schrödinger equation*

$$-\ddot{y}(t) + \varepsilon V(\theta + \omega t) y(t) - E y(t) = 0 , \qquad \text{(CQS)}$$

as well as the *time–discrete quasiperiodic Schrödinger equation*

$$-(u_{n+1} + u_{n-1}) + \varepsilon V(\theta + n\omega) u_n - E u_n = 0 \qquad \text{(DQS)}$$

are both examples of *linear quasiperiodic equations*.

In the *periodic case* $\omega = 2\pi \left(\frac{p_1}{q}, \ldots \frac{p_d}{q} \right)$, $\frac{p_i}{q} \in \mathbb{Q}$ for all $i = 1, \ldots, l$, according to Floquet theory, the time evolutions of the solutions of (DQS) is determined by the eigenvalues of the matrix

$$A_q(\theta) = \prod_{j=0}^{q-1} \begin{pmatrix} 0 & 1 \\ -1 & \varepsilon V(\theta + j\omega) - E \end{pmatrix} \in \mathrm{SL}\,(2, \mathbb{R}),$$

(for the (CQS), if $\Phi_t(\theta, E)$ denotes the monodromy matrix, the evolution is determined by $\Phi_q(\theta, E)$). Moreover one can make a change of variables which transforms the system to a constant coefficient system (this is the so-called *reducible* case, which was first considered by Lyapunov [NS]).

Not all quasiperiodic systems are reducible: indeed one has to overcome a small divisors problem in order to show that a quasiperiodic Schrödinger equation is reducible [DS], [MP]. For un up–to–date review we refer to [El2].

The problem of reducibility can be formulated more generally for any linear quasiperiodic *skew-product* system. Let G denote some matrix subgroup of $\mathrm{GL}\,(D, \mathbb{R})$, $\mathfrak{g}$ its Lie algebra. Let $A : \mathbb{T}^d \to G$ and $B : \mathbb{T}^d \to \mathfrak{g}$ be continuous functions. Then

$$(\theta, X) \mapsto (\theta + \omega, A(\theta)X),$$

$$\dot{X}(t) = B(\theta + t\omega)X,$$

are (respectively) discrete and continuous time skew-products. Of course two examples are given by (DQS) (with $G = \mathrm{SL}\,(2, \mathbb{R})$) and (CQS) (in this case $B \in \mathfrak{sl}\,(2, \mathbb{R})$). Again, a skew-product system will be reducible if there exists a change of variable which transforms it to a constant coefficient skew-product, i.e. A (resp. B in the continuous case) is constant.

Problem 4.1 (L.H. Eliasson) [El1, El2] *Show that any generic analytic one-parameter family of skew-systems sufficiently close to constant coefficients is reducible for almost every parameter value.*

A result very close to the previous problem has been obtained in [Kr] for general compact matrix groups.

4.2   Spectral theory and integrated density of states

An important notion in the study of quasiperiodic Schrödinger equations is the *rotation number* (or *integrated density of states*, see [JM]): for all E and for all initial data the following limit exists

$$\alpha(E) := \lim_{t \to \infty} \frac{1}{t} \arg(y(t) + i\dot{y}(t)), \quad \text{for the (CQS)}.$$

If the system is reducible then the imaginary parts of the eigenvalues of the constant coefficient matrix $B \in \mathfrak{sl}\,(2, \mathbb{R})$ to which it is transformed are $\pm\alpha(E)$.

When ε is small the properties of the rotation number as E varies are known: it is an absolutely continuous function. Generically it is constant outside a Cantor set. When ε^{-1} is small "localized" solutions can exist.

Problem 4.2.1 (H. Eliasson) *What are the properties of the map $E \mapsto \alpha(E)$ when ε^{-1} is small? Is it singular continuous? Absolutely continuous?*

Problem 4.2.2 (H. Eliasson) *Consider the (DQS) and suppose that there exist a measurable function $E_\infty(\alpha)$ and for a.e. α a function $\mathcal{U}_\alpha : \mathbb{T}^d \to \mathbb{R}$ ($\mathcal{U}_\alpha$ belongs to $L^2(\mathbb{T}^d)$ and is measurable in α) such that $u_n = e^{in\alpha}\mathcal{U}_\alpha(\theta+n\omega)$, $n \in \mathbb{Z}$, $E = E_\infty(\alpha)$ is a solution of (DQS). Assume that if for some subset Y of $\mathbb{T}$ one has $E_\infty(Y) = 0$ then the Haar measure of Y is also null. Is the spectrum purely absolutely continuous?*

4.3   Nonlinear Hamiltonian PDEs

Here we consider nonlinear Hamiltonian Partial Differential Equations (PDEs) in the finite-volume case. This means that we are concerned with equations for (vector) functions $u = u(x,t)$ where the space variable x belongs to some bounded domain (with Dirichlet or periodic boundary conditions). The idea is to treat Hamiltonian PDEs as ordinary differential equations in some infinite-dimensional function spaces formed by functions of the space variable x and to assume that they can be written in the form

$$\dot{u}(t) = J\nabla H(u(t)) \,, \tag{H}$$

where J is an anti self-adjoint operator in L^2 and ∇ denotes the L^2-gradient. Equation (H) is Hamiltonian w.r.t. the symplectic structure

$$\omega(v_1(x), v_2(x)) = \langle (-J)^{-1}v_1(x), v_2(x)\rangle_{L^2} \,.$$

A typical example is provided by the nonlinear Schrödinger equation

$$\dot{u} + i\Delta u - iu|u|^{2p} = 0 \tag{NLS}$$

with $x \in \mathbb{T}^n$, $p \in \mathbb{N}$. (NLS) can be rewritten in Hamiltonian form taking $J = i$ and $H(u) = \frac{1}{2}\int |\nabla u|^2 + \frac{1}{2p+2}\int |u|^{2p+2}$. Both $H(u)$ and the L^2-norm $\int |u|^2$ are preserved under the flow. The case $p = 1$ (cubic NLS) in one spatial dimension is integrable.

Another classical example of integrable Hamiltonian PDE is the Kortweg-de Vries (KdV) equation under zero-meanvalue periodic boundary conditions [GKM, No, La, DMN, MT]

$$\dot{u} = \frac{\partial}{\partial x}(-u_{xx} + 3u^2) \,, \quad x \in \mathbb{T}^1 \,, \quad \int_0^{2\pi} u(x,t)dx = 0 \,. \tag{KdV}$$

In this case integrability means the following: for all $n \in \mathbb{N}$ the space $\mathcal{H} = \cap_{s>0}H_0^s(\mathbb{T}^1)$ (here $H_0^s(\mathbb{T}^1)$ is the Sobolev space formed by zero-meanvalue functions on $\mathbb{T}^1$) contains a smooth invariant $2n$-dimensional manifold $\mathcal{T}^{2n}$ such that

(a) the restriction of (KdV) to $\mathcal{T}^{2n}$ defines a Liouville–Arnold integrable Hamiltonian system;

(b) $\mathcal{T}^{2n} \subset \mathcal{T}^{2m}$ if $m > n$;

(c) $\cup \mathcal{T}^{2n}$ is dense in each $H_0^s(\mathbb{T}^1)$.

Moreover the invariant manifolds $\mathcal{T}^{2n}$ are filled with time-quasiperiodic solutions (the so-called *n-gap solutions*, see the lectures of S. Kuksin in this volume).

Also the Sine–Gordon (SG) equation under Dirichlet boundary conditions

$$u_{tt} = u_{xx} - \sin u, \quad u(t, 0) = u(t, \pi) = 0, \tag{SG}$$

is integrable (i.e. it has *n*-gap solutions) but the manifolds $\mathcal{T}^{2n}$ have algebraic singularities and their union is proved to be dense only near the origin.

We refer to [B] and [Ku] for an introduction to the recent progress on the theory of nonlinear Hamiltonian PDEs.

Problem 4.3.1 (S. Kuksin) *Find a Lyapounov Theorem (not of KAM type) for Hamiltonian PDEs. I.e. prove that (under reasonable assumptions) most of time-periodic solutions from any non-degenerate one-parameter family persist under small Hamiltonian perturbations of the equation.*

For further information and some progress in this direction see [Ba2].

Problem 4.3.2 (S. Kuksin) *Find a non-perturbative way to obtain time-periodic solutions of an Hamiltonian PDE (the solutions should not be travelling waves).*

The main object of the Lectures of Kuksin in this volume is the proof of a KAM-type theorem wich assures the persistence under small Hamiltonian perturbations of most finite-gap solutions of Lax-integrable Hamiltonian PDEs like (KdV) or (SG). Nothing is known on the infinite-gap solutions (almost periodic solutions, see [AP] for an introduction). Following [MT] (see also [BKM]), one can ask:

Problem 4.3.3 (S. Kuksin) *Any Sobolev space $H_0^m(\mathbb{T}^1)$, $m \geq 1$, is filled by almost periodic solutions of (KdV). Do most of them persist under small Hamiltonian perturbations of the equation ?*

Presumably they do not.

From the KAM theorem described in Kuksin's lectures it also follows [BK] that most small amplitude finite-gap solutions of the φ^4 equation

$$u_{tt} = u_{xx} - mu + \gamma u^3, \quad m, \gamma > 0, \tag{φ^4}$$

persist under small Hamiltonian perturbations. The (φ^4) equation can be indeed obtained from the (SG) equation developing $\sin u$ into Taylor series at the third order and truncating. If one sets $m = 0$ into (φ^4) the existence of small-amplitude time-quasiperiodic solutions is not known.

Problem 4.3.4 (S. Kuksin) *Construct small amplitude time quasiperiodic solutions for the massless φ^4 equation $u_{tt} = u_{xx} - u^3$, $x \in \mathbb{T}^1$.*

Let us consider again the (NS) equation. As we have seen the time-flow preserves the Hamiltonian and the L^2-norm of the solutions. One can study the following properties of the long-time behaviour of the solutions:

(a) $\lim_{t \to \infty} \|u(t)\|_m = \infty$ (here m is sufficiently large);
(a') $\limsup_{t \to \infty} \|u(t)\|_m = \infty$;
(b) $u(t) \to 0$ weakly in H^s as $t \to \infty$;
(b') $u(t_n) \to 0$ weakly in H^s for a suitable sequence $t_n \to \infty$.

Problem 4.3.5 (S. Kuksin) *Which properties among a)–b') hold for the solutions of (NS) with a typical initial condition u_0, $u(0, x) = u_0(x) \in H^m$?*

Note the following facts:

- (a)–(b') are *all* wrong for *any* u_0 if $p = 1$, $n = 1$;
- (a)–(b') are wrong for *some* u_0 (for any p, n) since (NS) has time-periodic solutions;
- (b) $\Rightarrow$ (a) and (b') $\Rightarrow$ (a') due to the conservation of the L^2-norm.

Problem 4.3.6 (S. Kuksin) *Does (NS) have an invariant measure in H^s such that its support (in H^s) is a set with non-empty interior (here s is sufficiently large)?*

Not that a positive answer to Problem 4.9 implies that a) is wrong almost everywhere with respect to the measure.

Problem 4.3.7 (S. Kuksin) *In the spirit of Nekhoroshev's theorem one can ask the following: does an arbitrary solution of a ε-perturbed KdV equation remain ε^a-close to some finite gap torus during a time interval $0 \le t \le T_\varepsilon$ with $T_\varepsilon \ge C_M \varepsilon^{-M}$ for all $M > 0$?*

Probably a crucial step is to prove this statement for $T = \varepsilon^{-2}$. A first result in this direction has been obtained in [Ba1].

References

[Al]   V.M. Alekseev "Quasirandom oscillations and qualitative questions in celestial mechanics" Amer. Math. Soc. Transl. **116** (1981) 97–169

[AL]   S. Aubry and P.Y. Le Daeron "The discrete Frenkel–Kantorova model and its extensions I: exact results for the ground states" *Physica* **8D** (1983) 381–422

[AP]   L. Amerio and G. Prouse "Almost–Periodic Functions and Functional Equations" Van Nostrand Reinhold Company (1971)

[Ar]   V.I. Arnold "Small denominators and problems of stability of motion in classical and celestial mechanics" Russ. Math. Surv. **18** (1963) 85–191

[B]   J. Bourgain "Harmoinc Analysis and Nonlinear Partial Differential Equations" Proceedings of the ICM 1994, Birkhäuser (1995) Volume I 31–44

[Ba1]   D. Bambusi "Nekhoroshev theorem for small amplitude solutions in nonlinear Schrödinger operators" *Math. Z.* **230** (1999) 345–387

[Ba2]   D. Bambusi "Lyapunov center theorem for some nonlinear PDE's: a simple proof" *Ann. S.N.S. Pisa Cl. Sci.* **29** (2000) 823–837

[BG]   A. Berretti and G. Gentile, "Bryuno function and the standard map", University of Roma (Italy), Preprint (1998).

[BHS]   H. Broer, G.B. Huitema and M.B. Sevryuk "Quasi-periodic motions in families of dynamical systems" Springer Lect. Notes in Math. **1645** (1996)

[Bi]   G.D. Birkhoff "Dynamical Systems" Amer. Math. Soc., Providence, RI (1927)

[BK]   A.I. Bobenko and S.B. Kuksin "The nonlinear Klein–Gordon equation on an interval as a perturbed Sine–Gordon equation" Comment. Math. Helv. **70** (1995) 63–112

[BKM]   D. Bättig, T. Kappeler and B. Mityagin "On the Kortweg–de Vries equation: convergent Birkhoff normal form" *J. Funct. Anal.* **140** (1996) 335–358

[Bo]   J. B. Bost "Tores invariants des systèmes dynamiques hamiltoniens" Séminaire Bourbaki **639**, Astérisque **133–134** (1986) 113–157

[BPV]   N. Buric, I. Percival and F. Vivaldi "Critical Function and Modular Smoothing" *Nonlinearity* **3** (1990) 21–37

[Br]   A. D. Brjuno "Analytical form of differential equations" *Trans. Moscow Math. Soc.* **25** (1971) 131–288; **26** (1972) 199–239

[CL]   T. Carletti and J. Laskar "Scaling law in the standard map critical function. Interpolating Hamiltonian and frequency map analysis" *Nonlinearity* **13** (2000) 2033–2061

[CM]   T. Carletti and S. Marmi "Linearization of Analytic and Non-Analytic Germs of Diffeomorphisms of (ℂ, 0)" *Bull. Soc. Math. France* **128** (2000) 69–85

[Da]   A.M. Davie "The critical function for the semistandard map" *Nonlinearity* **7** (1994) 219–229

[De1]   A. Denjoy "Sur les courbes définies par les équations différentielles à la surface du tore" *J. Math. Pures et Appl.* **9** (1932) 333–375

[De2]   A. Denjoy "Les trajectoires à la surface du tore" *C. R. Acad. Sci. Paris* **223** (1946) 5–7

[DL]   D. De Latte "Diophantine conditions for the linearization of commuting holomorphic functions" *Discr. and Cont. Dynam. Sys.* **3** (1997) 317–332

[DMN]   B.A. Dubrovin, V.B. Matveev and S.P. Novikov "Nonlinear equations of Korteweg-de Vries type, finite zone linear operators and Abelian varieties" Russ. Math. Surv. **31:1** (1976) 55–136

[DS]   E.I. Dinaburg and Ya.G. Sinai "The one-dimensional Schródinger equation with quasi-periodic potential" Functional Anal. Appl. **9** (1975) 8–21

[El1]   L.H. Eliasson "Floquet solutions for the one-dimensional quasi-periodic Schrödinger equation" Commun. Math. Phys. **146** (1992) 447–482

[El2]   L.H. Eliasson "Reducibility and Point Spectrum for Linear Quasi-Periodic Skew-Products" Proceedings of the ICM 1998, Volume II, Doc. Math. J. DMV, 779–787

[Fo]   G. Forni "Analytic destruction of invariant circles" *Ergod. Th. and Dyn. Sys.* **14** (1994) 267–298

[GKM]   C.S. Gardner, M.D. Kruskal and R.M. Miura "Kortweg–de Vries equation and generalisations. II. Existence of conservation laws and constants of motion" *J. Math. Phys.* **9** (1968) 1204–1209

[Ha]   R.S. Hamilton "The Inverse Function Theorem of Nash and Moser" *Bull. A.M.S.* **7** (1982) 65–222

190   S. Marmi, J.-C. Yoccoz

[He1]   M. R. Herman "Résultats Récents sur la Conjugaison Différentiable" Pro-
          ceedings of the ICM 1978, Volume 2, 811–820
[He2]   M. R. Herman "Sur la conjugaison différentiable des difféomorphismes
          du cercle a des rotations" *Publ. Math. I.H.E.S.* **49** (1979) 5–234
[He3]   M. R. Herman "Sur les courbes invariantes par les difféomorphismes de
          l'anneau. Volume 1" *Astérisque* **103–104** (1983)
[He4]   M. R. Herman "Recent results and some open questions on Siegel's lin-
          earization theorem of germs of complex analytic diffeomorphisms of $\mathbf{C}^n$
          near a fixed point" Proc. VIII Int. Conf. Math. Phys. Mebkhout and
          Seneor eds. (Singapore: World Scientific) (1986), 138–184
[He5]   M. R. Herman "Sur les courbes invariantes par les difféomorphismes de
          l'anneau. Volume 2" *Astérisque* **144** (1986)
[He6]   M. R. Herman "Some Open Problems in Dynamical Systems" Proceed-
          ings of the ICM 1998, Volume II, *Doc. Math. J. DMV* 797–808
[HT]   G.R. Hall and M. Turpin "Robustness of periodic point free maps of the
          annulus" *Topology Appl.* **69** (1996) 211–215
[JM]   R.A. Johnson and J. Moser "The rotation number for almost periodic
          potentials" Commun. Math. Phys. **84** (1982) 403–438
[KH]   A. Katok and B. Hasselblatt "Introduction to the modern theory of dy-
          namical systems" Encyclopedia of Mathematics and its Applications **54**,
          Cambridge University Press, (1995)
[KO1]   Y. Katznelson and D. Ornstein "The differentiability of the conjugation
          of certain diffeomorphisms of the circle" *Ergod. Th. and Dyn. Sys.* **9**
          (1989) 643–680
[KO2]   Y. Katznelson and D. Ornstein "The absolute continuity of the conjuga-
          tion of certain diffeomorphisms of the circle" *Ergod. Th. and Dyn. Sys.*
          9 (1989) 681–690
[Kr]   R. Krikorian "Réducibilité des systèmes produits–croisés à valeurs dans
          des groupes compacts" Astérisque **259** (1999)
[Kra]   B. Kra "The conjugating map for commuting groups of circle diffeomor-
          phisms" *Israel Jour. Math.* **100** (1996) 303–316
[KS]   K.M. Khanin and Ya. Sinai "A new proof of M. Herman's theorem"
          Commun. Math. Phys. **112** (1987) 89–101
[Ku]   S.B. Kuksin "Elements of a Qualitative Theory of Hamiltonian PDEs"
          Proceedings of the ICM 1998, Volume II, Doc. Math. J. DMV, 819–829
[La]   P.D. Lax "Periodic solutions of the KdV equation" *Comm. Pure Appl.
          Math.* **28** (1975) 141–188
[M1]   J. Moser "Monotone twist mappings and the calculus of variations" *Er-
          god. Th. and Dyn. Sys.* **6** (1986) 401–413
[M2]   J. Moser "On commuting circle mappings and simultaneous diophantine
          approximations" *Math. Z.* **205** (1990) 105–121
[Ma1]   J.N. Mather "Existence of quasi-periodic orbits for twist homeomor-
          phisms of the annulus" *Topology* **21** (1982) 457–467
[Ma2]   J.N. Mather "More Denjoy invariant sets for area preserving diffeomor-
          phisms" *Comment. Math. Helv.* **60** (1985) 508–557
[Ma3]   J.N. Mather "Destruction of invariant circles" *Ergod. Th. and Dyn. Sys.*
          8 (1988) 199–214
[Ma4]   J.N. Mather "Variational construction of orbits for twist diffeomorphism-
          s" *Journal of the A.M.S.* **4** (1991) 203–267

[Ma5]   J.N. Mather "Action minimising invariant measures for positive definite Lagrangian systems" *Math. Z.* **207** (1991) 169–207

[Mar]   S. Marmi "Critical Functions for Complex Analytic Maps" *J. Phys. A: Math. Gen.* **23** (1990), 3447–3474

[MMY1]   S. Marmi, P. Moussa and J.-C. Yoccoz "The Brjuno functions and their regularity properties" *Commun. Math. Phys.* **186** (1997), 265–293.

[MMY2]   S. Marmi, P. Moussa and J.-C. Yoccoz "Complex Brjuno Functions" *Journal of the A.M.S.* **14** (2001) 783–841

[MP]   J. Moser and J. Pöschel "An extension of a result by Dinaburg and Sinai on quasi-periodic potentials" *Comment. Math. Helvetici* **59** (1984) 39–85

[MS]   S. Marmi and J. Stark "On the standard map critical function" *Nonlinearity* **5** (1992) 743–761

[MT]   H.P. McKean and E. Trubowitz "Hill's operator and Hyperelliptic function theory in the presence of infinitely many branch points" *Commun. Pure Appl. Math.* **29** (1976) 143–226

[No]   S.P. Novikov "The periodic problem for the Korteweg-de Vries equation" *Functional Anal. Appl.* **8** (1974) 236–246

[NS]   V.V. Nemytskii and V.V. Stepanov "Qualitative theory of differential equations" Princeton University Press (1960)

[PM]   R. Perez–Marco "Non linearizable holomorphic dynamics having an uncountable number of symmetries" *Invent. Math.* **119** (1995) 67–127

[S]   C.L. Siegel "Iteration of analytic functions" . *Ann. of Math.* **43** (1942), 807–812

[SK]   Ya.G. Sinai and K.M. Khanin "Smoothness of conjugacies of diffeomorphisms of the circle with rotations" *Russ. Math. Surv.* **44** (1989), 69–99

[St]   S. Sternberg "Infinite Lie groups and the formal aspects of dynamical systems" J. Math. Mech. **10** (1961), 451–474

[Yo1]   J.-C. Yoccoz "Conjugaison différentiable des difféomorphismes du cercle dont le nombre de rotation vérifie une condition diophantienne" *Ann. Sc. E.N.S.* **17** (1984) 333–359

[Yo2]   J.-C. Yoccoz "An introduction to small divisors problem", in "From number theory to physics", M. Waldschmidt, P. Moussa, J.-M. Luck and C. Itzykson (editors) Springer-Verlag (1992) pp. 659–679

[Yo3]   J.-C. Yoccoz "Théorème de Siegel, nombres de Bruno et polynômes quadratiques" *Astérisque* **231** (1995), 3–88.

[Yo4]   J.-C. Yoccoz "Centralisateurs et conjugaison différentiable des difféomorphismes du cercle" *Astérisque* **231** (1995), 89–242

[Yo5]   J.-C. Yoccoz "Analytic linearisation of circle diffeomorphisms", this volume

C.I.M.E. Session
on Dynamical Systems and Small Divisors
List of participants

F. ALESSIO, Dipartimento di Matematica, Politecnico di Torino, Corso Duca degli Abruzzi, 24, 10129 Torino, Italia

B. FAYAD, Pennsylvania State University, Department of Mathematics, 413 Mcallister Building, University Park, PA 16802, U.S.A

P. BERNARD, 121, rue de Turenne, 75003 Paris, France

A. BERRETTI, Dipartimento di Matematica, Università di Roma Tor Vergata, Via della Ricerca Scientifica, 00133 Roma, Italia

U. BESSI, Scuola Normale Superiore, Piazza dei Cavalieri, 7, 56126 Pisa, Italia

D. BOSIO, School of Mathematical Sciences, Queen Mary and Westfield College, Mile End, London E1 4 NS, UK

T. CARLETTI, Dipartimento di Matematica U. Dini, Viale Morgagni, 67/a, 50134 Firenze, Italia

C. CARMINATI, Dipartimento di Matematica, Via Buonarroti, 2, 56127 Pisa, Italia

D. CORTICELLI, Scuola Normale Superiore, Piazza dei Cavalieri, 7 56126 Pisa, Italia

S. CROVISIER, ANS-Lyon Chb B53, 46, allée d'Italie, 69364 Lyon Cedex 07, France

D. DE LATTE, Mathematics, University of North Texas, PO Box 305118, Denton, TX 76203-5118, USA

P. DUCLOS, CPT CNRS, Luminy case 907, 13288 Marseille cedex 9, France

R. FABBRI, Dipartimento di Matematica, Viale Morgagni, 67/a, 50134 Firenze, Italia

J. FEJOZ, 13, rue d'Arcole, 75004 Paris, France

G. GENTILE, Dipartimento di Matematica, Università di Roma Tre, Largo San Leonardo Murialdo, 1, 00146 Roma, Italia

C. GIBERTI, Dipartimento di Matematica pura ed applicata, Via Campi, 213/B, 41100 Modena, Italia

T. GRAMCHEV, Dipartimento di Matematica, Via Ospedale; 72, 09124 Cagliari, Italia

D. GUZZETTI, SISSA, Via Beirut, 2-4, 34014 Trieste, Italia

H. ITO, Department of Mathematics, Tokyo Institute of Technology, Okayama, Meguro-ku, Tokyo, 152-8551, Japan

R. KRIKORIAN, Centre de Mathématiques, Ecole Polytechnique, URA-CNRS 169, 91128 Palaiseau, France

U. LOCATELLI, Observatoire de la Cote d'Azur, B.P. 4229, 06304 Nice Cedex, France

M. MACRÍ, Dipartimento di Matematica, Università di Napoli, Napoli, Italia

M. MAZZOCCO, SISSA, Via Beirut, 2–4, 34014 Trieste, Italia

P. MONTECCHIARI, Dipartimento di Matematica, Piazzale Europa, 1, 34127 Trieste, Italia

P. MOUSSA, Service de Physique Theorique, CEA-Saclay, 91191 Gif-sur-Yvette Cedex, France

Y. OKUYAMA, Department of Mathematics, Graduate School of Science, Kyoto, Japan

A. M. SANZ GIL, Menendez Pelayo, N 12, 3 I 47001 Valladolid, Spain

D. SAUZIN, CNRS, Bureau des Longitudes, 77, av. Denfert-Rochereau, 75014 Paris, France

S. SENTI, Université de Paris-Sud, Paris XI, Laboratoire de Topologie et Dynamique, Mathematiques, Bâtiment 425, 91405 Orsay Cedex, France

S. SMIRNOV, Max-Planck-Instiut für Mathematik, Gottfried-Claren-Str., 53225 Bonn, Germany

L. STOLOVITCH, Université P. Sabatier, Laboratoire E. Picard, 118, route de Narbonne, 31062 Toulouse Cedex 4, France

M. STRIANESE, Via Salv. Calenda n. 6/H, 84126 Salerno, Italia

C. VERNIA, Dipartimento di Matematica pura ed applicata, Via Campi, 213/B, 41100 Modena, Italia

M. VITTOT, Center for Theoretical Physics, Case 907 - CNRS - Luminy, 13288 Marseille cedex 9, France

F. WAGENER, Department of Mathematics, University of Groningen, P.O. Box 800, 9700 AV Groningen, The Netherlands

C. ZHANG, Departement et Laboratoire de Mathematiques, La Rochelle, France

LIST OF C.I.M.E. SEMINARS

1954   1. Analisi funzionale   C.I.M.E
       2. Quadratura delle superficie e questioni connesse   "
       3. Equazioni differenziali non lineari   "

1955   4. Teorema di Riemann-Roch e questioni connesse   "
       5. Teoria dei numeri   "
       6. Topologia   "
       7. Teorie non linearizzate in elasticità, idrodinamica, aerodinamic   "
       8. Geometria proiettivo-differenziale

                                                                  "

1956   9. Equazioni alle derivate parziali a caratteristiche reali   "
     10. Propagazione delle onde elettromagnetiche   "
     11. Teoria della funzioni di più variabili complesse e delle funzioni   "
          automorfe

1957 12. Geometria aritmetica e algebrica (2 vol.)   "
     13. Integrali singolari e questioni connesse   "
     14. Teoria della turbolenza (2 vol.)   "

1958 15. Vedute e problemi attuali in relatività generale   "
     16. Problemi di geometria differenziale in grande   "
     17. Il principio di minimo e le sue applicazioni alle equazioni   "
          funzionali

1959 18. Induzione e statistica   "
     19. Teoria algebrica dei meccanismi automatici (2 vol.)   "
     20. Gruppi, anelli di Lie e teoria della coomologia   "

1960 21. Sistemi dinamici e teoremi ergodici   "
     22. Forme differenziali e loro integrali   "

1961 23. Geometria del calcolo delle variazioni (2 vol.)   "
     24. Teoria delle distribuzioni   "
     25. Onde superficiali   "

1962 26. Topologia differenziale   "
     27. Autovalori e autosoluzioni   "
     28. Magnetofluidodinamica   "

1963 29. Equazioni differenziali astratte   "
     30. Funzioni e varietà complesse   "
     31. Proprietà di media e teoremi di confronto in Fisica Matematica   "

1964 32. Relatività generale   "
     33. Dinamica dei gas rarefatti   "
     34. Alcune questioni di analisi numerica   "
     35. Equazioni differenziali non lineari   "

1965 36. Non-linear continuum theories   "
     37. Some aspects of ring theory   "
     38. Mathematical optimization in economics   "

196

1966	39. Calculus of variations	Ed. Cremonese, Firenze
	40. Economia matematica	"
	41. Classi caratteristiche e questioni connesse	"
	42. Some aspects of diffusion theory	
1967	43. Modern questions of celestial mechanics	"
	44. Numerical analysis of partial differential equations	"
	45. Geometry of homogeneous bounded domains	
1968	46. Controllability and observability	"
	47. Pseudo-differential operators	"
	48. Aspects of mathematical logic	
1969	49. Potential theory	"
	50. Non-linear continuum theories in mechanics and physics and	"
	their applications	
	51. Questions of algebraic varieties	"
1970	52. Relativistic fluid dynamics	"
	53. Theory of group representations and Fourier analysis	"
	54. Functional equations and inequalities	"
	55. Problems in non-linear analysis	"
1971	56. Stereodynamics	"
	57. Constructive aspects of functional analysis (2 vol.)	"
	58. Categories and commutative algebra	"
1972	59. Non-linear mechanics	"
	60. Finite geometric structures and their applications	"
	61. Geometric measure theory and minimal surfaces	
1973	62. Complex analysis	"
	63. New variational techniques in mathematical physics	"
	64. Spectral analysis	"
1974	65. Stability problems	"
	66. Singularities of analytic spaces	"
	67. Eigenvalues of non linear problems	
1975	68. Theoretical computer sciences	"
	69. Model theory and applications	
	70. Differential operators and manifolds	"
1976	71. Statistical Mechanics	Ed. Liguori, Napoli
	72. Hyperbolicity	"
	73. Differential topology	
1977	74. Materials with memory	"
	75. Pseudodifferential operators with applications	"
	76. Algebraic surfaces	"
1978	77. Stochastic differential equations	Ed. Liguori, Napoli
	78. Dynamical systems	&
		Birkhäuser
1979	79. Recursion theory and computational complexity	"
	80. Mathematics of biology	"

1980	81. Wave propagation		Ed. Liguori, Napoli
	82. Harmonic analysis and group representations		&
	83. Matroid theory and its applications		Birkhäuser
1981	84. Kinetic Theories and the Boltzmann Equation	(LNM 1048)	Springer-Verlag
	85. Algebraic Threefolds	(LNM 947)	"
	86. Nonlinear Filtering and Stochastic Control	(LNM 972)	"
1982	87. Invariant Theory (LNM 996)		"
	88. Thermodynamics and Constitutive Equations	(LN Physics 228)	"
	89. Fluid Dynamics	(LNM 1047)	"
1983	90. Complete Intersections	(LNM 1092)	"
	91. Bifurcation Theory and Applications	(LNM 1057)	"
	92. Numerical Methods in Fluid Dynamics	(LNM 1127)	"
1984	93. Harmonic Mappings and Minimal Immersions	(LNM 1161)	"
	94. Schrödinger Operators	(LNM 1159)	"
	95. Buildings and the Geometry of Diagrams	(LNM 1181)	"
1985	96. Probability and Analysis	(LNM 1206)	"
	97. Some Problems in Nonlinear Diffusion	(LNM 1224)	"
	98. Theory of Moduli	(LNM 1337)	"
1986	99. Inverse Problems	(LNM 1225)	"
	100.Mathematical Economics	(LNM 1330)	"
	101. Combinatorial Optimization	(LNM 1403)	"
1987	102. Relativistic Fluid Dynamics	(LNM 1385)	"
	103. Topics in Calculus of Variations	(LNM 1365)	"
1988	104. Logic and Computer Science	(LNM 1429)	"
	105.Global Geometry and Mathematical Physics	(LNM 1451)	"
1989	106. Methods of nonconvex analysis	(LNM 1446)	"
	107.Microlocal Analysis and Applications	(LNM 1495)	"
1990	108.Geometric Topology: Recent Developments	(LNM 1504)	"
	109.H∞ Control Theory	(LNM 1496)	"
	110. Mathematical Modelling of Industrial Processes	(LNM 1521)	"
1991	111. Topological Methods for Ordinary	(LNM 1537)	"
	Differential Equations		
	112. Arithmetic Algebraic Geometry	(LNM 1553)	"
	113. Transition to Chaos in Classical and	(LNM 1589)	"
	Quantum Mechanics		
1992	114. Dirichlet Forms	(LNM 1563)	"
	115. D-Modules, Representation Theory,	(LNM 1565)	"
	and Quantum Groups		
	116. Nonequilibrium Problems in Many-Particle	(LNM 1551)	"
	Systems		
1993	117. Integrable Systems and Quantum Groups	(LNM 1620)	"
	118. Algebraic Cycles and Hodge Theory	(LNM 1594)	"
	119. Phase Transitions and Hysteresis	(LNM 1584)	"

1994	120. Recent Mathematical Methods in Nonlinear Wave Propagation	(LNM 1640)	Springer-Verlag
	121. Dynamical Systems	(LNM 1609)	"
	122. Transcendental Methods in Algebraic Geometry	(LNM 1646)	
1995	123. Probabilistic Models for Nonlinear PDE's	(LNM 1627)	"
	124. Viscosity Solutions and Applications	(LNM 1660)	"
	125. Vector Bundles on Curves. New Directions	(LNM 1649)	"
1996	126. Integral Geometry, Radon Transforms and Complex Analysis	(LNM 1684)	"
	127. Calculus of Variations and Geometric Evolution Problems	(LNM 1713)	
	128. Financial Mathematics	(LNM 1656)	"
1997	129. Mathematics Inspired by Biology	(LNM 1714)	"
	130. Advanced Numerical Approximation of Nonlinear Hyperbolic Equations	(LNM 1697)	"
	131. Arithmetic Theory of Elliptic Curves	(LNM 1716)	"
	132. Quantum Cohomology	(LNM 1776)	"
1998	133. Optimal Shape Design	(LNM 1740)	"
	134. Dynamical Systems and Small Divisors	(LNM 1784)	"
	135. Mathematical Problems in Semiconductor Physics	to appear	"
	136. Stochastic PDE's and Kolmogorov Equations in Infinite Dimension	(LNM 1715)	"
	137. Filtration in Porous Media and Industrial Applications	(LNM 1734)	"
1999	138. Computational Mathematics driven by Industrial Applications	(LNM 1739)	"
	139. Iwahori-Hecke Algebras and Representation Theory	to appear	"
	140. Theory and Applications of Hamiltonian Dynamics	to appear	"
	141. Global Theory of Minimal Surfaces in Flat Spaces	(LNM 1775)	"
	142. Direct and Inverse Methods in Solving Nonlinear Evolution Equations	to appear	"
2000	143. Dynamical Systems	to appear	"
	144. Diophantine Approximation	to appear	"
	145. Mathematical Aspects of Evolving Interfaces	to appear	"
	146. Mathematical Methods for Protein Structure	to appear	"
	147. Noncommutative Geometry	to appear	
2001	148. Topological Fluid Mechanics	to appear	
	149. Spatial Stochastic Processes	to appear	
	150. Optimal Transportation and Applications	to appear	
	151. Multiscale Problems and Methods in Numerical Simulations	to appear	

Fondazione C.I.M.E.

Centro Internazionale Matematico Estivo
International Mathematical Summer Center
http://www.math.unifi.it/~cime
cime@math.unifi.it

2002 COURSES LIST

Real Methods in Complex and CR Geometry

June 30 - July 6 - Martina Franca (Taranto)

Course Directors:

Prof. Dmitri Zaitsev (Università di Padova), zaitsev@math.unipd.it
Prof. Giuseppe Zampieri (Università di Padova), zampieri@math.unipd.it

Analytic Number Theory

July, 10-19 - Cetraro (Cosenza)

Course Directors:

Prof. C. Viola (Università di Pisa), viola@dm.unipi.it
Prof. A. Perelli (Università di Genova), perelli@dima.unige.it

Imaging

September, 15 - 21 - Martina Franca (Taranto)

Course Directors:

Prof. George Papanicolaou (Stanford University), papanico@georgep.Stanford.edu,
Prof. Giorgio Talenti (Università di Firenze), talenti@math.unifi.it

4. Lecture Notes are printed by photo-offset from the master-copy delivered in camera-ready form by the authors. Springer-Verlag provides technical instructions for the preparation of manuscripts. Macro packages in T_EX, L^AT_EX2e, L^AT_EX2.09 are available from Springer's web-pages at

http://www.springer.de/math/authors/b-tex.html.

Careful preparation of the manuscripts will help keep production time short and ensure satisfactory appearance of the finished book.

The actual production of a Lecture Notes volume takes approximately 12 weeks.

5. Authors receive a total of 50 free copies of their volume, but no royalties. They are entitled to a discount of 33.3 % on the price of Springer books purchase for their personal use, if ordering directly from Springer-Verlag.

Commitment to publish is made by letter of intent rather than by signing a formal contract. Springer-Verlag secures the copyright for each volume. Authors are free to reuse material contained in their LNM volumes in later publications: A brief written (or e-mail) request for formal permission is sufficient.

Addresses:

Professor Jean-Michel Morel
CMLA, École Normale Supérieure de Cachan
61 Avenue du Président Wilson
94235 Cachan Cedex France
e-mail: Jean-Michel.Morel@cmla.ens-cachan.fr

Professor Bernard Teissier
Institut de Mathématiques de Jussieu
Equipe "Géométrie et Dynamique"
175 rue du Chevaleret
75013 PARIS
e-mail: Teissier@ens.fr

Professor F. Takens, Mathematisch Instituut
Rijksuniversiteit Groningen, Postbus 800
9700 AV Groningen, The Netherlands
e-mail: F.Takens@math.rug.nl

Springer-Verlag, Mathematics Editorial, Tiergartenstr. 17
D-69121 Heidelberg, Germany
Tel.: +49 (6221) 487-701
Fax: +49 (6221) 487-355
e-mail: lnm@Springer.de